W0016700

Conspicuous Gallantry

CIVIL WAR IN THE NORTH

Conspicuous Gallantry

The Civil War and
Reconstruction Letters of
James W. King,
11th Michigan Volunteer
Infantry

Edited by Eric R. Faust

The Kent State University Press
Kent, Ohio

Material from the James W. King Collection appears courtesy of Western Michigan University Archives and Regional History Collections. Material from the Benjamin F. Bornder papers, the James Martin letters, the Wells Family papers, and the Anson De Puy Van Buren papers appear courtesy of the Bentley Historical Library, University of Michigan

© 2015 by The Kent State University Press, Kent, Ohio 44242

Library of Congress Catalog Card Number 2014049078
ISBN 978-1-60635-243-4
Manufactured in the United States of America

Library of Congress Cataloging-in-Publication Data
King, James W., 1842–1903.
Conspicuous gallantry : the Civil War and Reconstruction letters of James W. King, 11th Michigan Volunteer Infantry / edited by Eric R. Faust.
pages cm. — (Civil War in the North)
Includes bibliographical references and index.
ISBN 978-1-60635-243-4 (hardcover : alkaline paper) ∞
1. King, James W., 1842–1903—Correspondence.
2. United States. Army. Michigan Infantry Regiment, 11th (1861–1864)
3. Soldiers—Michigan—Correspondence.
4. Michigan—History—Civil War, 1861–1865—Personal narratives.
5. United States—History—Civil War, 1861–1865—Personal narratives.
6. United States—History—Civil War, 1861–1865—Regimental histories.
7. Cotton farmers—Southern States—Correspondence.
8. Reconstruction (U.S. history, 1865–1877)—Tennessee—Sources.
9. Reconstruction (U.S. history, 1865–1877)—Alabama—Sources.
10. Newspaper editors—Michigan—Correspondence.
I. Faust, Eric R., 1971– II. Title.
E514.511TH .K56 2015
973.7'474—dc23
2014049078

19 18 17 16 15   5 4 3 2 1

For Sandra,
who always believes in me just a little more
than I do in myself.

And now you have before you one of the most startling episodes of the war; I cannot render it in words; dictionaries are beggarly things. But I may tell you they did not storm that mountain as you would think. They dash out a little way, and then slacken; they creep up, hand over hand, loading and firing, and wavering and halting, from the first line of works to the second; they burst into a charge with a cheer, and go over it. Sheets of flame baptize them; plunging shot tear away comrades on left and right; it is no longer shoulder to shoulder; it is God for us all! Under tree trunks, among rocks, stumbling over the dead, struggling with the living; facing the steady fire of eight thousand infantry poured down upon their heads as if it were the old historic curse from heaven, they wrestle with the Ridge. Ten, fifteen, twenty minutes go by like a reluctant century. The batteries roll like a drum; between the second and last lines of rebel works is the torrid zone of the battle; the hill sways up like a wall before them at an angle of forty-five degrees, but our brave mountaineers are clambering steadily on—up— upward still! You may think it strange, but I would not have recalled those men if I could. They would have lifted you, as they did me, in full view of the heroic grandeur; they seemed to be spurning the dull earth under their feet, and going up to do Homeric battle with the greater gods.

—*Journalist Benjamin Franklin Taylor,*
eyewitness to the Battle of Missionary Ridge

Contents

———✦———

Acknowledgments

The process of writing this book has ranked among the most rewarding undertakings of my life. This was a journey I could never have completed alone, and numerous people have earned my heartfelt gratitude for helping me along the way.

First, thanks to Sharon Carlson and the entire staff at Western Michigan University Archives and Regional History Collections for allowing me the opportunity and honor of publishing the James W. King Collection.

Next, I would like to recognize everyone at Kent State University Press for making the entire publishing process a pleasure. I am particularly indebted to Joyce Harrison, who recognized the merit in these letters at a time when the manuscript was at best a diamond in the rough. Without her support and encouragement, this project might never have seen the light of day.

Brian Craig Miller and Timothy J. Orr pored over the text and offered meticulous feedback and invaluable suggestions for improvement. Both of these gentlemen perceived value and significance in this letter collection from angles I had not even considered. Thank you both for enabling me to do justice to the topic. Copyeditor Margery Tippie was a pleasure to work with as well and saved me from numerous potential embarrassments. Any remaining errors and omissions are solely my responsibility.

Many archivists and librarians deserve mention here as well. Karen Jania and the staff of the Bentley Historical Library are to be commended for their efficiency and professional courtesy in ensuring convenient access to Bentley's expansive and indispensable materials. Kevin Driedger at the Library of Michigan was instrumental in providing access to resources vital to understanding James King's career as a journalist. The staff of the Albuquerque/Bernalillo County Library cheerfully processed a mountain of interlibrary loan requests. John Allison at the Morgan County (Alabama) Archives shared his wealth of knowledge about

Decatur in the immediate aftermath of the war and pointed me to sources that eased the detective work required to make some sense out of the Kings' harrowing, and poorly documented, experiences down South in 1867.

A big thank-you to noted Civil War cartographer Hal Jespersen, whose stunning maps grace this book. It is my fervent hope, Hal, to send more business your way soon. Thanks for not laughing when you saw my own abortive attempts at map making.

Researcher Vonnie Zullo was extremely helpful in obtaining materials from the National Archives. Thanks to her knowledge and experience, her archive visits proved just as fruitful as if I had been there myself.

I was not, by a far cry, the only King descendant who was aware of, and fascinated by, these letters. But for a twist of fate, this volume would likely have been written by my cousin, John Dudd. My thanks to his wife, Joan, for generously sharing his notes and transcriptions with me. Three other cousins— Rebecca Shank and Michael and Howard King—joined me for an immensely rewarding genealogy rap session one day at the Three Rivers Public Library, and helped to ensure that I had my facts straight regarding the Kings of the nineteenth century.

In such a massive undertaking, the degree of support received from one's family and friends can make or break the entire endeavor. Our children, Adrian and Nina, cheerfully put up with their father disappearing now and then to visit some archive or battlefield in a faraway land. An occasional pat on the back, or its verbal equivalent, sustained my efforts long before I was fortunate enough to obtain the feedback and encouragement of the publisher and peer reviewers. My mother, Priscilla Camarillo, my brother, Trent, and my wonderful wife, Aleksandra—with her unbounded, unconditional supportiveness—all perused early drafts of the manuscript and shared in the joys of the countless discoveries that illuminated my path.

Historical Cast

———∞∞∞———

Individuals mentioned more than once in James W. King's letters are listed here to ease the reader's burden in keeping names straight. In most cases, additional biographical details are provided later in a note associated with the initial mention of each person in a letter. Additional listings are provided for King and Babcock family members, for soldiers of the 11th Michigan whose writings are frequently cited, and for select people who fall short of the above criteria but earn further mention in the editor's supporting comments. Major historical figures (e.g., Lincoln, Grant, Sherman, Lee, and Davis) are excluded here. Two separate lists are provided: the first enumerates people referenced informally (without a surname), while the second lists individuals by last name.

INDIVIDUALS REFERENCED BY FIRST NAME, MIDDLE NAME, OR NICKNAME

Alden: See Fulkerson, Loriston Alden.
Aunty: Freedwoman domestic servant on the Critz plantation.
Billy: In letters of 1861 only, Billy is an unidentified individual. For 1862–1863 see Davis, William H.; for 1864 and later see Iddings, William Clarence.
Crockett: Freedman field hand on the Critz plantation.
Darius: See Babcock, Darius Ambrose Jr.
Dick: See Dixon, Cuthbert.
Dio: See Babcock, Darius Ambrose Jr.
Dix/Dixie: See Dixon, Cuthbert.
Etta/Ettie/Etty: See Babcock, Mary Esther.
Fie: See Iddings, Fiene.
Harriet: Freedwoman domestic servant on the Critz plantation.
Jane/Jenny: See Babcock, Sarah Jane.

J. Webster: See King, Jennie May. (This was her nickname before she was born—James assumed the baby was a boy.)

Kate: Freedwoman field hand on the Critz plantation.

Lem/Lemuel: See Pierce, Lemuel Packard.

Let/Letitia: See Dunn, Eliza Letitia.

Linus: See Squire, Linus Truman.

May: See King, Jennie May.

Odd: See Davis, William H.

Rebecca: See King, Martha Rebecca.

Tommy: See Iddings, Thomas.

Vet: Unidentified individual interested in purchasing James's farmland in Fabius, Michigan.

Webb: See Holbrook, Daniel Webster.

Will: See Iddings, William Clarence.

Listing by surname

Babcock, Darius Ambrose, Jr. ("Dio"): Jenny's brother.

Babcock, Darius Ambrose, Sr.: Jenny's father.

Babcock, James Frank: Jenny's brother.

Babcock, John Simeon: Jenny's brother.

Babcock, Mary Esther ("Etta," "Ettie," or "Etty"): Jenny's sister.

Babcock, Perry Bruce: Jenny's brother.

Babcock, Ruth Butler (Blodgett): Jenny's mother.

Babcock, Ruth Rozanna: Jenny's sister.

Babcock, Sarah Jane ("Jane," "Jenny"): James's sweetheart and future wife.

Baur, Mr.: Unidentified individual interested in purchasing James's farmland in Fabius, Michigan.

Bennett, Benjamin Grove: Captain, Company D. Commissioned major January 7, 1863. Killed at Missionary Ridge November 25, 1863.

Blair, Austin: Governor of Michigan.

Blodgett, John Wesley: Jenny's uncle.

Bordner, Benjamin F.: Sergeant, Company D. Letters preserved at Bentley Historical Library.

Bournes, William: Private, Company C. Native Irishman and friend of the Babcocks.

Bragg, Braxton: Commanding general of the Confederate Army of Tennessee.

Brown, Neill Smith: Ex-governor of Tennessee.

Buell, Don Carlos: Union general. Commanded the Department of the Ohio.

Crittenden, Thomas Leonidas: Union general. Commanded Rosecrans's Left Wing at Stones River and the 21st Corps at Chickamauga.

Critz, Jacob, Jr.: Owner of a cotton plantation in Thompson's Station, Tennessee.

Davis, William H. ("Billy," "Odd"): Private, Company A. Friend of the Babcocks.

Dixon, Cuthbert ("Dick," "Dix," or "Dixie"): Sergeant, Company A. Promoted to first sergeant March 21, 1863; commissioned second lieutenant April 19, 1864. Wounded near Dallas, Georgia, May 27, 1864.

Drake, Addison T.: Quartermaster. Acting assistant division quartermaster from February 1863; acting assistant corps quartermaster February through August 1864. Commissioned captain, Company K, June 17, 1864.

Dunn, Eliza Letitia ("Let," "Letitia"): Jenny's first cousin.

Elliott, William N.: Surgeon.

Ewing, Andrew: Former congressman. Judge on the Army of Tennessee's military court.

Floyd, John Buchanan: Former secretary of war. Confederate general.

Fulkerson, Loriston Alden: Jenny's first cousin. Private, Company A. Died of disease February 15, 1862.

Garfield, James Abram: Former Ohio state senator and future U.S. president. Appointed Rosecrans's chief of staff in spring 1863.

Gillaspie, Ira: Private, Company C. Promoted to corporal August 10, 1862. Diary published in 1965.

Hall, Ephraim Gaylord: First sergeant, Company I. Commissioned second lieutenant, Company F, April 1, 1862; first lieutenant August 18, 1862. Wounded and taken prisoner at Stones River December 31, 1862. Commissioned captain February 19, 1863. Acting brigade assistant adjutant general June 17, 1864.

Harkness, Lindley R.: Private, Company E, enlisted August 9, 1862. Promoted to commissary sergeant March 1863.

Hart, Benjamin F.: Sergeant, Company D. Commissioned first lieutenant January 9, 1864.

Hicks, Borden Mills: Sergeant, Company E. Commissioned second lieutenant November 15, 1862; captain September 20, 1863.

Hoisington, William W.: First sergeant, Company E. Died of typhoid August 15, 1862.

Holbrook, Daniel Webster ("Webb"): Private, Company A. Wounded near Murfreesboro January 23, 1863. Discharged for disability October 28, 1863.

Iddings, Fiene ("Fie"): William Iddings's wife.

Iddings, Thomas ("Tommy"): William Iddings's brother.

Iddings, William Clarence ("Will"): Private, Company A.

Johnson, Andrew: Former governor of Tennessee, appointed military governor March 1863. Lincoln's second-term vice president and his successor as president.

Keegan, Patrick H.: First lieutenant, Company K. Commissioned captain January 1, 1863; major November 25, 1863. Wounded at Missionary Ridge November 25, 1863.

Kies, Lewis G.: Clerk in Ezra Benham Kirk's quartermaster office.

King, Benjamin Montaigne: James's father.

King, Eliza (Van Buren): James's stepmother.

King, Eunice: James's sister.

King, Eva Fay: James and Jenny's daughter.

King, Henry: James's brother.

King, Henry Burr: James and Jenny's son.

King, Herbert Holbrook: James and Jenny's son.

King, James Guy: James and Jenny's son.

King, Jennie May ("J. Webster," "May"): James and Jenny's firstborn.

King, John: James's brother. Enlisted as private, Company G, 96th Illinois Infantry, September 6, 1862. Wounded at Lookout Mountain November 24, 1863.

King, John Willard: James and Jenny's son.

King, Martha Rebecca ("Rebecca"): James's sister.

King, Martha (Wetherbee): James's biological mother.

Kinkade, Nancy: Kentuckian plantation lessee in Thompson's Station, Tennessee.

Kirk, Ezra Benham: Assistant quartermaster, regular army.

Liddle, Byron J.: Private, Company D. Captured at Gallatin August 13, 1862. Killed at Ruff's Station July 4, 1864.

Martin, James: Corporal, Company H. Promoted to sergeant November 13, 1862; first sergeant September 1, 1863. Letters preserved at Bentley Historical Library.

May, William J.: Colonel. Resigned April 1, 1862.

McCook, Alexander: Union general. Commanded Rosecrans's Right Wing at Stones River, and the 20th Corps at Chickamauga.

Morgan, John Hunt: Confederate cavalryman famous for raids behind Union lines.

Mudge, Melvin B.: Captain, Company B. Commissioned lieutenant colonel January 7, 1863. Wounded at Chickamauga, September 20, 1863.

Negley, James Scott: Union general. Commanded the 11th Michigan's division at Stones River and Chickamauga.

Oakes, David Jr.: Captain, Company A. Died of typhoid January 30, 1863.

Pierce, Lemuel Packard ("Lem"): Jenny's first cousin. Corporal, Company A. Division train master April 1863. Discharged for disability May 6, 1864.

Price, Mark B.: Clerk in Ezra Benham Kirk's quartermaster office.

Raymond, William E.: Private, Company F, enlisted December 6, 1861. Appointed hospital steward December 25, 1863. Letters preserved at Bentley Historical Library.

Rice, Charles: James's childhood friend. Regimental band member. Enlisted in regular brigade band August 4, 1863. Wounded near Atlanta July 26, 1864.

Rose, Daniel: Private, Company A. Diary and letters preserved at Western Michigan University Archives and Regional History Collections.

Rosecrans, William Starke: Union general. Commanded the Army of the Cumberland at Stones River and Chickamauga.

Rousseau, Lovell Harrison: Union general. Commanded a division at Stones River and Chickamauga.

Sill, Joshua Woodrow: Union general. Killed leading a brigade at Stones River.

Snyder, Washington Irving: Corporal, Company E. Promoted to sergeant January 1, 1863; sergeant major April 7, 1863. Died October 5, 1863, of wounds received at Chickamauga September 20, 1863.

Spencer, Anson: Eliza King's nephew. Private, Company A. Wounded at Stones River December 31, 1862. Discharged for disability May 28, 1863.

Spencer, Ezra: Eliza King's nephew. Corporal, Company E. Promoted to sergeant. Killed at Stones River December 31, 1862.

Squire, Linus Truman: Sergeant, Company H. Commissioned second lieutenant June 24, 1862; first lieutenant January 1, 1863. Acting adjutant February 1863. Commissioned adjutant August 3, 1863.

Stanley, Timothy Robbins: Union colonel. Commanded the 11th Michigan's brigade at Stones River and Chickamauga.

Stoughton, William Lewis: Lieutenant colonel. Commissioned colonel April 1, 1862. Wounded at Ruff's Station July 4, 1864. Brevetted major general March 13, 1865.

Sturges, Aaron B.: First sergeant, Company A. Commissioned second lieutenant February 7, 1862. Resigned due to disability February 13, 1863.

Sutton, Ms. E.: Schoolteacher. Possibly Elizabeth Sutton of Lockport, Michigan.

Thomas, George Henry: Union general. Commanded Rosecrans's Center Wing at Stones River, the 14th Corps at Chickamauga, and the Army of the Cumberland afterward, continuing through the Atlanta Campaign.

Underwood, John H.: Quartermaster sergeant. Commissioned first lieutenant, Company G, November 14, 1862. Acting quartermaster February 1863 through September 1864.

Van Buren, Anson De Puy: James's uncle; educator and author of *Jottings of a Year's Sojourn in the South.*

Wells, Benjamin Franklin: Private, Company A. Promoted to corporal January 24, 1862; sergeant March 21, 1863. Letters preserved at Bentley Historical Library.

Whallon, James M.: Sergeant, Company C. Promoted to sergeant major March 12, 1862; commissioned second lieutenant January 7, 1863. Resigned due to disability January 15, 1864.

Whelan, Arvin F.: Commissioned assistant surgeon November 12, 1861. Resigned October 13, 1862.

Whitney, William G.: Sergeant, Company B. Commissioned second lieutenant January 7, 1863; first lieutenant June 17, 1864. Wounded in action at Chickamauga. Awarded Medal of Honor October 21, 1895, for gallantry at Chickamauga.

Introduction

⎯⎯◦∞◦⎯⎯

Long after the last shots of the Civil War were fired, James Wood King looked back on that defining period of his life and described himself as just another one of the countless young farm boys who had gone off to war. His military experiences exemplified those of many Northern soldiers. He enlisted partly out of a simple sense of duty to his country, but he was also driven by the belief that only a strong and whole Union could uplift its citizens' quality of life. Like many of his peers, he took up arms in the firm conviction that he must defend, and thus prove himself worthy of, the free country his ancestors had fought to create in the Revolutionary War. Throughout his three-year enlistment, disease, bullets, and shells threatened to exact from James King the ultimate price for his ideals, yet he never wavered. Rather, he strove to match the courage of his comrades—and in one momentous hour eclipsed their valor with his own. But when it all began, the "irrepressible conflict" of slavery was little on his mind. In 1861 it was all about duty and the Union, and he called it "one of the noblest causes that mankind were ever engaged in."[1]

Although James King in some ways played the role of the everyman Union soldier, in many other aspects his experiences, his writings, and the story of his life stand out. In 1901 Senator Julius Caesar Burrows would declare him "numbered among our best citizens in Michigan." When James passed away two years later, a congressman would deliver his eulogy, the neighboring county's bar association would motion to attend his funeral en masse, and an ex-comrade would exclaim that no braver man had ever lived. On many different levels, the story of how this particular farm boy experienced the Civil War and its aftermath, and how he rose from humble beginnings to prominence in his home state, is a tale worth telling.

What, exactly, justifies the publication of James W. King's letters? The Civil War generation was quite literate, and the relative maturation of the postal system by that time allowed for millions of soldiers to write and send many millions of letters home. Enough of this correspondence survives and sees publication that it is difficult to keep up with the pace at which it appears. This letter collection, however, is exceptional. First, there is James's eloquence; he received a superior education, by nineteenth-century standards, and demonstrated it through a deep affinity for the choicest poetry and prose. This literary prowess infuses his correspondence with his sweetheart, Sarah Jane Babcock (Jenny), with a degree of romance beyond what is commonly seen in Civil War letters. Additionally, this Michigander's writings articulately portray the dual life of the soldier—the daily monotonies of military routine, starkly contrasted with the infrequent but nightmarish realities of combat. His pen equally illuminates the manner in which Union soldiers' views on race evolved during the war. Furthermore, James King's letters offer the reader a glimpse of wartime Michigan, which has received little scrutiny in comparison with the eastern states of the Union, and, more specifically, grants insight into James's regiment, the hard-fighting 11th Michigan Volunteer Infantry, a unit worthy of more scholarly attention than it has received.

Another standout feature of James King's story is his unusually well-documented example of rank-and-file heroics, an act that resulted in a serious wound with life-

James Wood King. (Author's collection)

long personal consequences. But perhaps most important, the appeal of James King's letters does not stop with his muster out, or even with the end of the war. His postwar correspondence bequeaths to posterity a firsthand account that defies the long-held stereotype of the carpetbaggers—supposedly lowlife Yankees who were presumed to have craved political power and plundered the South of its capital after the war—a stereotype that historians, in recent decades, have only begun to dispel.[2] James's postwar experience then proceeds to illustrate the struggles of a disabled soldier transitioning to peacetime and shares a glimpse into the American political scene from Reconstruction through the early years of the Gilded Age, when he served as editor of a major Republican newspaper.

Where did James King acquire such a superior education? The surviving writings of Civil War soldiers from Michigan demonstrate that the young state had established a sound system of public education in the antebellum period. Yet there is a singular quality about James's letters. Whether seated in a Sibley tent or crouched down in a trench, he effortlessly quoted the likes of Sir Walter Scott, Lord Byron, and a host of other literary giants, leveraging some of the most expressive words ever penned in order to elucidate the sights, travails, and traumas of camp life and battle. Evidence suggests that the primary driving force behind James's extraordinary schooling was his uncle, Anson De Puy Van Buren, a prolific author and highly prominent figure in the early history of education in Michigan. Van Buren's twenty-one-year career as an educator began at a log cabin school in Battle Creek, Michigan, in 1838—the year after Michigan achieved statehood—and continued at institutions of higher learning almost to the beginning of the Civil War.

In 1857 Van Buren sought a warmer climate for health reasons and directed an academy near Yazoo City, Mississippi, for one year. He returned to Michigan and authored the book *Jottings of a Year's Sojourn in the South*, a popular Northern account of life in Dixie that was well received in both sections. James memorized significant portions of his uncle's volume, including sections that quoted the great poets in much the same style James would exhibit in his letters. A comparison of the writings of Anson Van Buren with the postwar words of James King reveals similarities in their politics as well, particularly in their desire for reform.[3] There can be little doubt that the younger man was influenced by his elder.

James King and Sarah Jane Babcock's courtship via correspondence presents another interesting aspect of these letters, offering insight into the evolutionary state of relationships in the 1860s. With the advent of the war, how would millions of lovers cope with multi-year long-distance relationships in an era when almost all interpersonal contact was face to face? Beginning with James's

Sarah Jane Babcock.
(Author's collection)

letters of 1861, we witness the early stages of his and Jenny's relationship and observe its gradual progression toward marriage. In the process, we are introduced to customs foreign to the modern reader. For example, the very act of falling in love was still gaining acceptance as a socially tolerated phase in finding a mate. Eighteenth-century dogma had held that marriage was a practical arrangement, and assigning priority to romantic desire when selecting a partner might still, even in the mid-nineteenth century, be considered immature and irresponsible. James clearly reveled in falling in love, yet the expression of his affection is strikingly repressed in his initial correspondence with Jenny—the flowery prose and poetry came later—and both of the young lovers repeatedly expressed fears over anyone else seeing their letters. Even after marriage, James asked Jenny whether the intensity of his love for her was "wicked," and he begged her time and again not to let another soul view the intimate words he penned for her eyes.[4]

Even in America's deadliest war, most soldiers spent only a tiny fraction of their enlistment period engaged in combat. James's letters relate the daily routine of Civil War soldiering—standing guard, cooking, laundering, filling out paperwork, distributing supplies, enduring chronic illness, anxiously awaiting the next mail, and bearing with comrades' profanity, gambling, and drunkenness—with no less attention to detail than he employed in describing the horrors of traversing a battlefield. His words impart to the reader a comprehension of how the Union soldier experienced both life and death.[5]

White Northerners' views of blacks gradually evolved as the war dragged on, and for James King that evolution proceeded through twists and turns throughout the war years and beyond. Racism was such an overwhelming norm in contemporary society that even James, the son of an avowed antislavery man, looked upon blacks during the war as markedly inferior, and he exhibited attitudes toward slaves and freedmen that ranged from dismissiveness to condescension. After the human cost of the war skyrocketed and emancipation became a reality, this same soldier, who had formerly looked down upon those in bondage, stoutly defended the proclamation that granted their freedom. Then, after the war's end, while supervising ex-slave labor in Tennessee, James's perception of the freedmen regressed to the point of contempt—only to rise again, after years of more mature reflection, and give birth to a firm and publicly declared belief in the full equality of all men.[6]

Much of Michigan's Civil War story remains untold. At the outbreak of the war, Michigan had been a state for only twenty-four years. Her treasury was empty, and the state's entire military presence consisted of about 1,200 militia, who were later described by the state's wartime adjutant general, John Robertson, as being "rather looked upon as a burlesque on the military profession." Poorly equipped, ill trained, and wholly unprepared to meet an emergency of the magnitude soon to be encountered, the militiamen nonetheless represented the seeds from which sprang thirty-one regiments of infantry—thirty white and one black—eleven regiments of cavalry, fourteen artillery batteries, a regiment of sharpshooters, and one of engineers. The state, with an 1860 population of just over 700,000, sent 90,000 of its sons off to war between 1861 and 1865. Almost 15,000 of them laid down their lives. Far more ink has been spilled about the Civil War in the East than in the West, and correspondingly, the Union states of what is now called the Midwest have been the subject of far less study than those of the Eastern Seaboard.[7] James King's regiment, the 11th Michigan Infantry, has lain under the same cloak of obscurity as the state that spawned it. The 11th fought tenaciously at Stones River, at Chickamauga, at Missionary Ridge, and in the Atlanta Campaign, yet it remains to this day a unit of unsung heroes. After the war, many of the 11th's veterans spoke and wrote of their comrades' historic deeds, but none proved to be the bard who would secure the unit's legacy. The regiment's accomplishments remain largely unrecognized to this day.[8]

Next, there is the matter of James King's conspicuous gallantry at the Battle of Missionary Ridge, a striking example among the innumerable acts of rank-and-file

James Wood King in 1864. (Courtesy Western
Michigan University Archives and Regional
History Collections)

bravery performed by soldiers on both sides during the Civil War. James, joining
in a battle completely outside his line of duty—in an act that would earn him a
nomination for the Medal of Honor—twice charged ahead of his brigade against
fortified enemy lines, launching himself into a melee in the first case, and in the
second instance becoming one of the first soldiers to exploit a gap that triggered
the rout of an entire enemy division. In the process, he received a crippling wound
that would forever alter the course of his civilian life.

Of especial interest among James King's letters is the correspondence he
penned during his stint as a carpetbagger in Thompson's Station, Tennessee,
in 1866. James headed south to raise cotton after the war and inevitably found
himself entangled in the great social and political upheavals and controversies
of the postwar South. The story of his first season of cotton cultivation in
Tennessee defies the timeworn portrayal of carpetbaggers as an infestation of
immoral Yankees who bilked Dixie of its wealth. James loved Tennessee, got
along effortlessly with the locals—ex-Confederates or not—sank his life savings
into the local economy, and devoutly wished to settle down there. But his plans
in Tennessee did not pan out financially, and his decision to make another go
at cotton farming near Decatur, Alabama, in the following year proved fateful:
the well-meaning Kings ran afoul of the Ku Klux Klan and were driven out
with devastating monetary loss.

James King's rich postwar biography provides an informative glimpse into
the struggles faced by disabled soldiers trying to transition themselves back into
the civilian world and offers a unique viewpoint into the politics of the ensuing
decades.[9] After years of struggling to get by financially—with his prewar occupation of farming now a physical impossibility—a marvelous stroke of luck

leveraged James's talent for shorthand writing into gainful employment as editor of the *Lansing Republican,* one of Michigan's premiere political newspapers. His political voice and opinions are preserved in newspaper print, granting us insight into the political climate of Michigan, and of the nation, in the 1870s and 1880s.

Here, then, is the correspondence of James Wood King. All but one of the letters published herein are preserved in the James W. King Collection at the Western Michigan University Archives and Regional History Collections; the sole exception, the letter of March 30, 1863, resides with the Anson De Puy Van Buren Papers at University of Michigan's Bentley Historical Library. When quoting from other historical sources, I have retained the original spelling, grammar, capitalization, and punctuation. In transcribing the letters, I have selectively edited minor errors with the intention of minimizing distractions for the reader. On occasion I removed passages that are repetitious or contain obscure references that could not be deciphered through additional research. But without exception, top priority has been granted to the accurate presentation of the author's thoughts.

CHAPTER 1

Drum and Fife

September–December 1861

———✺———

It is because I loved my home so dear
that I left it.

In May 1832 Benjamin Montaigne King, accompanied by his pregnant wife, Martha, and toddler son, Henry, completed a six-hundred-mile trek from New York State to Michigan Territory. The young pioneer family had journeyed west to embrace the adventure and opportunity of the frontier. Benjamin was intercepted by soldiers at Jonesville, Michigan, and drafted to serve one month in the Black Hawk War. Martha spent a seeming eternity isolated in a log cabin with her young son, expecting hostile Indians to overrun her little home at any moment. Benjamin returned from his mercifully uneventful conscription unscathed, and the emergency passed. The Kings settled down at the future site of Eschol, and Benjamin rented a room to labor as a cobbler, pursuing the trade he had apprenticed for as a teen in New York. In 1834 the Kings relocated to Three Rivers, where they occupied the first house built in that town. In the following year, Benjamin built a new home for his family, and a year later the native New Yorker purchased from the government 120 acres of land southwest of Three Rivers. For years he spent the days clearing farmland while devoting his evenings to shoemaking.[1]

The Kings were descended from Clement King, an Englishman who crossed the ocean and became constable of Marshfield, Massachusetts, in 1668. Also counted among the King ancestors was Johannes de La Montanye, a French doctor born about 1595 and educated in medicine at Leiden University. Montanye later settled in America and was appointed vice director of the colony of New Netherland in 1656.[2]

Benjamin and Martha raised their family in the near-wilderness of southwestern Michigan. John was born in September 1832; Eunice in 1840. On August 13, 1842, Martha gave birth to James Wood King. Two years later, the birth of James's sister Martha Rebecca, as fate would have it, completed the family. Martha, the pretty wife who had followed her husband deep into the frontier, died in 1846, shy of her fortieth birthday. Two years after that, Benjamin remarried, to Eliza Van Buren, a cousin of President Martin Van Buren.[3]

In 1859 Benjamin had a brick house constructed on his farm, which was now part of the township of Fabius. James King's friend and former schoolmate, Charles Rice, performed the construction, while James hauled the brick and mortar.[4] James had worked summers on the family farm, starting around the age of ten, and attended school during winters through the age of sixteen, at which time he became an assistant teacher at the school in nearby Lockport. "He spent hours on Webster's Elementary Speller," noted a biographical album of St. Joseph County, "where his companions spent minutes. It was the same in penmanship, reading, arithmetic and grammar . . . which enabled him to become thoroughly familiar with whatever subject he had at hand." A determined young man, James was described by one friend as "one of the last to quit in any contest."[5]

James grew close to schoolmate Sarah Jane Babcock ("Jenny" to her friends), and the two teens fell in love even as their country began to disintegrate over the issue of slavery. Jenny, born on December 12, 1844, was the daughter of Darius Ambrose and Ruth Butler (Blodgett) Babcock, who immigrated to Michigan by 1840. In addition to four living siblings, Jenny was blessed with about eighty first cousins on the prolific Blodgett side of the family. Three of her great-grandfathers and two of her great-great-grandfathers had fought against the crown in the Revolutionary War, and one of these soldiers from each side of her family held a seat in Connecticut's House of Representatives in 1781.[6] Her grandfather Ezra Blodgett was a descendant of Mayflower Pilgrims John and Priscilla (Mullins) Alden. Through Ezra's wife, Clarissa, Jenny was descended from England's King Edward III.[7]

When the Confederates fired on Fort Sumter on April 12, 1861, a wave of indignation and patriotism swept the North, and Michigan was no exception. Benjamin King by this time was a known antislavery man, and when recruiting rallies sprang up throughout St. Joseph County, his son James enlisted in the St. Joseph County Guard. James surely went off to war mindful of his father's namesake, their Revolutionary War ancestor Benjamin Montanye. Montanye, who served as a post rider for George Washington, had ridden into capture near the end of the Revolution while carrying correspondence that was purportedly

Darius Ambrose Babcock.
(Author's collection)

Ruth Butler (Blodgett) Babcock. (Author's
collection)

spurious and intended for British eyes—specifically, letters crafted by General
Washington to deceive the British regarding his intentions for the military
campaign of 1781. Montanye's descendants, including the Kings, maintained
that their forefather's deception helped secure independence by enabling the
decisive victory at Yorktown. To James, his revered great-grandfather was a role
model whose example proved that one brave man could impact the course of
a great war.[8]

The federal and state governments were unprepared for the overwhelming
response they received to the call for volunteers, and as a result it was not until
August 24, 1861, shortly after James's nineteenth birthday, that he and the rest of
his nascent military unit—destined to become Company A of the 11th Michigan
Volunteer Infantry—gathered at the courthouse in Centreville, Michigan, to
muster into what was initially known as Colonel May's Independent Regiment.[9]
The raw recruits proceeded to the regimental rendezvous point, Camp Tilden,
at the outskirts of White Pigeon.[10] A respected lawyer from Sturgis by the
name of William Lewis Stoughton drilled the men. The regiment voted for its
officers and unanimously selected William J. May as colonel, with Stoughton
appointed lieutenant colonel.[11]

On September 8 Private James King, enjoying a brief furlough at home,
sat down and penned his first letter to Sarah Jane Babcock. His initial correspondence exudes the enthusiasm of a raw recruit experiencing military life
for the first time, yet something else is conspicuously absent. Though James

and Jenny had professed their love for each other less than one year prior, there is little hint of the young romance at first. Subsequent letters reveal that they were both fearful that others might read their correspondence. As the weeks pass, James's tone relaxes, and it gradually becomes apparent that there is more going on here than just friendship.

———————

Fabius, Michigan,
September 8, 1861

Dear Jenny, you may perhaps think strange of my not coming to see you as I agreed when I last saw you, but circumstances would not permit. I probably shall return once or twice again, and then I will try to put my promise in execution. So far a soldier's life is better than I expected, but I expect the worst is to come. Well, let it come, I am prepared for it. I do not suppose I can tell you anything about our doings at White Pigeon that would interest you. Probably Odd has related the whole narrative.[12] Our fare is not any of the best, but it will not do to grumble. We may get a great deal worse. Men who enlist cannot expect to find home in camp. If they do, they are very much mistaken. I am going to White Pigeon this morning. About sixty of the Three Rivers company go with us.[13] I have many acquaintances in that company, many who are good boys, and it will be quite a pleasure to have them with us. The camp has many attractions, but I think I should prefer the steady quiet of home if I could take my choice. I cannot write any more at present, but remember an ever true friend. JK.

William Lewis Stoughton.
(Gracie, *The Truth about Chickamauga*)

Camp Tilden,
White Pigeon, Michigan,
September 17, 1861

Dear Jenny, we are not to leave White Pigeon [for] several days. Upon our arrival here, we found that orders had been given for eighty of our men to volunteer [to leave the regiment], or if they did not, the whole regiment was to [disband], but before our arrival the number had volunteered to go.[14] They left us amid cheers and hurrahs. There were none among those who left that you knew. I found Lem [Jenny's cousin, Lemuel Pierce] in camp.[15] He was out when we left. I have not time to write any more at present. The time for drilling is near at hand, and no one must be absent. Excuse poor writing as our conveniences for writing are poor. But I begin to learn to put up with inconveniences. I think if we do stay here a week [that] I will get a chance to come home, but if I do not, remember him who will ever prove true to you. James W. King.

Camp Tilden,
White Pigeon, Michigan,
September 22, 1861

Dear Jenny, it is the Sabbath morning. I have just returned from breakfast, and having nothing to busy myself at, I thought I might as well write, for perhaps you would like to know what is going on in our camp. The drum and fife is now playing to call the guards together. Every morning, each company is called up and eight or ten men detailed out of each company, then the drum beats to call them together. After all assemble, they are divided into reliefs or companies, with sixteen in a company. These are numbered 1st, 2nd, and 3rd reliefs. Then the first relief is placed on duty. They stay on guard two hours, when they are relieved by the second relief, and so on for twenty-four hours.[16] This is the day of writing in our camp. There are four persons writing at the same desk. I do not know who they are writing to, but probably their sweethearts, as they take a great deal of pains and stop to think pretty often. I suppose Lemuel feels better. He has been rather down all the week until he found he could go home, when his spirits revived. Going home had a wondrous effect on him. I should like to have had it tried on me. I heard the col. say yesterday every man should have a chance of going home once more before we leave. If I can get away next Saturday, I am coming home. I received a letter from W. S. Wood. Heard he is encamped near Washington.[17] There has been no fighting yet, only a few picket skirmishes. He said the secesh had possession of a peach orchard near

David Oakes Jr. (Courtesy Family of Esther Thompson Hull)

the encampment. He took a notion to have some peaches. The boys tried to persuade him not to go, but he thought that peaches were deserving only to the brave, so he mustered around and found two of his companions who volunteered to go with him. They started out, keeping a good lookout for secesh. Upon arriving at the orchard they saw 3 secesh under a tree. They saw Will and his party, and fired at them. Will ordered his boys to lie down and they shot over him, then rising up, they fired on them and drove them out of the orchard, then they filled their haversacks with the delicious fruit and returned to camp uninjured. He says part of the time they live on the fat of the land, part on the lean, but mostly on the lean. I have not time to write any more at present. I probably shall see you again, but if I should not, success be with you is the prayer of him who thinks of you often. James King.

If you direct to White Pigeon in care of Capt. Oakes, write next week for it may be I shall not be home.[18]

Camp Tilden,
White Pigeon, Michigan,
September 29, 1861

Dear Jenny, it is afternoon and most of our boys have gone to service. There is to be preaching out on the parade ground in front of the quarters, but having some writing to do, and it being rainy and disagreeable out of doors, I preferred to remain in quarters. We have as yet received no orders to leave here. Gov. Blair is coming here on Tuesday.[19] I do not know his purpose, but [he] probably will try to get us in the state's service if we suit him. All of the men who are off on furlough are to be called in by Monday noon on account of his coming. The boys have just come in, and there is now a general rush. All is noise and bustle, I can hardly hear myself think. I left home Thursday morning, remaining at home on Wednesday. I came here on the ten o'clock train and found a deserted camp. Most of our boys had gone to the fair at Centreville. There were hardly enough of our company left for their complement of guards, and of course I had to stand guard. Lem thought he would play up smart, and volunteered to stand with me. We went on Friday morning and remained until Saturday morning at 9 o'clock. About 8 o'clock in the evening, it commenced raining. He thought if he had another chance to volunteer, he would not volunteer at all. Standing guard was not what it was cracked up to be, but fortunately it did not rain very hard, and he felt better. He is playing his tricks on everyone he meets. He has just gone to see if Dixie [Cuthbert Dixon] as he calls him is cheating the boys who is playing cards with him.[20] I suppose you may think we have rather a hard set here, and I guess you think right. I think if you would have heard all the oaths I have since I have been writing here, I am afraid you would have a poor opinion of camp life.[21] We have had a great many visitors here today, many of them I knew. Among them was C. P. Buck of Lockport, you may perhaps have heard of the gentleman.[22] I have just been reading a piece called sunshine at home, and now I should like to read a piece of sunshine in camp, and I think a letter from you would be it. I have not time to write any more as the bugle is now sounding for dress parade, so goodbye Jenny for this time.

Camp Tilden,
White Pigeon,
Michigan, October 4, 1861

Dear Jenny, I have received several letters this week, but yours is the choicest of them all. I should like to have been with you and your mother when you were alone, but that was among the impossibilities. All furloughs were rescinded on Saturday and none given out in expectation that Gov. Blair would be here on

Tuesday, but unhappily we were disappointed. Lieut. Col. Stoughton received a letter stating that he would be here on Thursday. Accordingly, at 9 o'clock people began to flock in, and by 2 o'clock in the afternoon quite a large crowd had assembled in expectation of seeing the gov. and also to see us drill. 2 o'clock came and with it the train, but no gov. Here we were doomed to disappointment again.[23] Miss E. Sutton, her sister, and brother-in-law were here yesterday among the rest who came to see us.[24] I did not get a chance to talk much with her. I saw her while in rank and shook hands with her. While at supper, she and her sister came in where we were eating. I asked her what she thought of soldiers' fare. She thought I could judge better than she. They left last night on the 10 o'clock train for Sturgis. We have as yet received no orders to leave. I cannot tell whether I will get a chance of coming home again or not. If we stay a week or two, we probably will get a chance of coming. I shall improve the opportunity if offered. I am enjoying myself better than I expected. While I am not on drill, I have plenty of reading and writing to do. We have a singing book here, and every evening we have a sing, our capt. joining with us. I have not heard from home since I left, but I think of it often. If anyone wishes to appreciate a good home, let him be placed in such a situation as I and a great many others. I do not wish to have you think that I repine at my lot. Oh no, it is because I loved my home so dear that I left it. I have not time to write any more at present. It is pretty near drill time, and all must be on hand. Give my best respects to your father and mother, and believe me ever yours. James.

> Camp Tilden,
> White Pigeon, Michigan,
> October 18, 1861

Dear Jenny, I received your letter a few moments ago and now sit down to answer, but I cannot accede to your proposal. Even If I wished to get discharged, I could not. The Captain would not let me off, he has appointed me clerk.[25] He has discharged several of his men, but said yesterday he should keep me anyway. No Jenny, I think it just as much my duty to go now as when I first enlisted. You said you did not wish me to let money tempt me.[26] I hope you did not have any such thought. Do you think I would have left home, friends and everything I held dear for sordid dross? No, it would pain me if I thought you could even think so. While your happiness is as dear to me as life, duty prompts me to go. My country first, home and friends next. Jenny, what would friends be to me if I had no country? I hope after this when you think of me, you will remember me as one who is doing his duty. I cannot help but admire your wishes for my safety, but after all it would be among the impossibilities for me to get a discharge.[27] Billy is here now and talks of enlisting, but I am going to try to talk him out of it.[28] I do not think he knows what he is going into. He cannot get in our company even if he wished to, it is full. I filled 3 muster rolls yesterday for Captain Oakes. He said he could not take any more. I have been writing most of the time for the last 3 days. I like it better than drilling. When I have no writing to do, I drill but have no guard duty to perform. I have done this writing in the eating house, up in his room. It is a pleasant place. I have one of the windows raised, and I have a fine view of the prairie and village. I wish you could see the prairie from this window as I saw it yesterday. [There is] nothing that I admire so much as a fine landscape view. Your hearing that we had received our uniforms and going to leave next week is false. We have received no uniforms or orders to march, but we look for the uniforms tonight.[29] There is some ladies coming to visit us. I do not know what they think of our house, but they look as though there was something to see. There are 3 of them, they stand near the desk watching Dixon and I. He is writing a letter. I cannot tell whether I can get home again or not. If I can I will, but if I should not see you again, may God bless and protect is the earnest wishes of him who will never forget you. James.

Camp Tilden,
White Pigeon,
Michigan, October 22, 1861

Dear Jenny, you may perhaps think me cruel and unjust, but I could not even if I wished to. I have been duly enlisted and sworn in, and my officers would not consent to it on no condition. I would do anything in my power to make you happy, but consider whether it would conduce to your happiness to have me remain at home while our country calls for volunteers. I think, Jenny, if you will take a second thought, you will think different. I am seated all alone in my quarters, silence reigns around. I got excused from drill this afternoon as I had some clothing lists to make out for Captain Oakes. I finished them a few moments ago, and I thought I would converse with you a few moments through the medium of the pen. We have received part of our uniforms, but not the whole. There was some mistake made in the shipment of the pants. There were only one hundred and sixty pairs sent on, when 1000 pairs were ordered. We have as yet received no marching orders, but expect to leave as soon as we get our equipment. If you get a chance of coming down here, come by all means. I should like to see you before I leave, and get that likeness. I cannot tell whether I will get a chance of coming home again or not, but I probably shall yet. We cannot tell today what the morrow will bring forth, but let us hope for the best when we do leave. I think we shall be ordered to Kentucky, as the 9th, 10th, and Eleventh Michigan Infantry is ordered there, and we are numbered as the Eleventh.[30] Last Sunday we were called upon to perform a task for the first time. That was of consigning a fellow soldier to the tomb. His name was G. F. Grather.[31] He enlisted with us at Centreville to share in the dangers and privations of a camp life, but alas, he has fallen before he met the enemies of his country. He was buried in the White Pigeon burying ground. His was the largest funeral I ever attended. There were nearly one thousand soldiers, beside about the same number of citizens. All seemed to feel the mournfulness of the occasion. All of the officers as [well] as the privates attended. I have been considerable busy this week, have not drilled much but have had a great deal of writing. I had rather do this than drill, and then another thing, I do not have to stand guard which is quite an item. I have not time to write you any more at present as I hear the boys coming, and when they get here you might as well give up at once, so goodbye Jenny, may God bless and protect you. Give my best respects to your father and mother. Write soon. James.

> Camp Tilden,
> White Pigeon,
> Michigan, October 27, 1861

Dear Jenny, although I cannot be with you today, yet it is some consolation to converse with you. Although I am seated in my quarters, my thoughts are not here, they wander away to different scenes from these. Strange is it not? Perhaps you can guess where they wander. The morning sun shines brightly through the open door of the quarters. Its glancing rays fall brightly on the home of the soldier. The busy hum of voices reach my ear, and the forms of soldiers meet my gaze on every side. Some are seated together talking, others writing, while some are thinking. Now what are they thinking of—'tis home. I can read it in their very looks. Home, what a world of thought arises at the mention of this word. A thousand recollections rush upon the mind, and memory goes rushing back among those dear old times that is now numbered among the past. There is something in this word that fascinates which none can tell.

> And none who came and went away
> Could tell wherein the fascination lay[32]

But Jenny I will leave off these meditations and talk to you on the scenes and doings in camp. We have been drilling for the past week, and distributing uniforms. There is only a part of them here, there were only enough come to dress the noncommissioned officers.[33] I will not describe it as you have likely seen Lemuel, who left here on Friday. I should have come home with him, but Capt. Oakes could not spare me. If the clothing would have come I should have had to make out clothing rolls, which is quite a job. The capt. told me he thought I would have a chance of getting home once more. We have received no orders as yet to march, but I expect we will leave for Detroit this week or next, where we will join the Ninth which is now quartered there, where we will get our arms and then be pushed on to Kentucky. I have been looking for a letter from you this week, but have been disappointed. It cannot be that you have forgotten me or think less of me for doing my duty. I cannot think it so—but I should like to know your thoughts on the subject. The dying words of David Crockett I have always kept in view. 'Tis this:

> This maxim keep—when I am dead
> See first that you are right then go ahead

Jenny I think I am right—do you not?
    Remember to write, and believe me ever yours. James.
The regiment is now lettered, and we have received the first letter. Perhaps when you write you had better put the number of the regiment on, which is called the Mich. Eleventh.
Odd lies in the bunk at sleep, all the rest of the boys are at dinner.

Camp Tilden,
White Pigeon,
Michigan, November 6, 1861

    Dear Jenny, I received your letters last night and am truly grateful to you for them. I had looked anxiously all the week but had been disappointed until last night, when judge of my surprise at receiving two from you, more than I expected. Your sentiment you showed merits a dozen letters. I knew, Jenny, if you would take a second thought, you would think different. I am proud to love one of so much force of character, for fifty others I know would have given way instead of bearing up as you have done. The reason I have not been home is on account of my having to help the capt. in distributing clothing, besides doing a great deal of writing. But I probably shall get a chance soon of coming home once more. I gave my place as clerk to D. W. [Daniel Webster] Holbrook, who was clerk at first.[34] He left our company on being offered a lieutenancy in another company, but the company broke up and he came back. I told him if he

Benjamin Franklin Wells. (*History of St. Joseph County, Michigan*)

wanted the place again I would not stand in his way, as he had been so kind to speak to the captain about my taking his place when he left. He said if I wished to keep my place he would not ask it, but I thought one good turn deserves another, and gave up. We have just received news of the capture of Floyd by Rosecrans with about seven thousand prisoners, but it needs confirmation.[35] A few more such blows would put a damper on secesh. The letter you spoke of as Odd's seeing was none of yours, I can assure you, as no one has ever seen them except my humble self. I done with that letter as you requested. But I should like to have kept it, had it not been your request to have it destroyed. As for my thinking you silly Jenny, I do not. A great many would have done worse. We have not drilled much this week on account of the weather, but the afternoon is very pleasant and I expect every minute to hear the order to fall in for drill. The order came, and I had to drop my letter and go and drill. We had a good time. Everything passed off nicely. The place where we drill is on a fine meadow lying east of the town. Every afternoon we are formed here for battalion drill. A thousand soldiers makes quite a show. I wish you could see us when we are on duty. There is a large crowd of people assembled every afternoon to see us drill. Among them are many I know. Almost every day I have some friends to see me, but there is one dearer to me than them all. Yes dear Jenny, I am happiest when I can stroll off by myself and think of you. I often wonder (wonder how that sounds, well I hope I am not the only person who wonders. I believe it used to be practiced pretty generally once upon a time) how you are enjoying yourself,

James's letter to Jenny, November 6, 1861. (Courtesy Western Michigan University Archives and Regional History Collections)

William W. Hoisington.
(*History of St. Joseph County, Michigan*)

and if you think of the many good times that is now numbered among the past. If the present does look dark, and heavy clouds hang over us, remember Jenny that darkness always gives way to light. "There is a better time a coming and I am sure 'tis coming here." I should like to have been with you Sunday evening. I am sorry that you were disappointed, but I thought I could only see you once more before I left, and I would wait until just before we left. I think I shall come sometime next week. I am glad to learn you are to commence going to school before long. I hope you will have a good teacher and will enjoy yourself. I should think it would be more pleasant for you to attend on the prairie, but likely you know best. Lem is lying near me asleep. Odd is spinning a yarn to Dixon and a young fellow by the name of [William Clarence] Iddings.[36] B. F. [Benjamin Franklin] Wells is writing on part of the desk.[37] The boys all seem to be enjoying themselves well. Billy, I am sorry to tell you, is lying in his quarters intoxicated. It is strange what beasts men will make of themselves. It was for this reason I tried to persuade him not to enlist. 'Tis a pity that one so young should be led away, but 'tis no more than I see every day—but let that pass. I must bring my letter to a close. You must not think me vexed on receiving two letters from you at a time. On the contrary, I could have read a dozen, for you do not know the joy your letters bring to me. You must excuse all mistakes and poor writing, as our conveniences for writing are very limited. The light is getting dim and I must close, so goodbye dear Jenny for this time, and believe me ever yours. James.

Camp Tilden,
White Pigeon,
Michigan, November 21, 1861

Dear Jenny, I arrived in camp Monday. Dixon, Wells, and [William W.] Hoisington accompanied me.[38] We got here about noon. Just in time to save our dinners. I found things in camp very near the same as when I left, only some of the boys had received more of their clothing, which made them look rather odd. Our clothing is not all here yet. The colonel is now in Detroit hurrying it up. When that comes, then probably we will go. On Sunday night, or rather Monday morning when I got home, I found things left in rather a careless manner; chairs were where they had ought not to be, besides a trunk which was left carelessly at the head of the stairs. But I was looking out for breakers, and happily surmounted all difficulties without any accidents. So much for their pains. Alden [Jenny's cousin, Loriston Alden Fulkerson] received your letter last night and is now answering it.[39] I expect you will get something better this time, as he takes great pains with it. Lemuel is complaining of being sick this morning. He was vaccinated last week and has a very bad arm.[40] Dick [Dixon] is lying in his bunk, he was out last night to a party up to Centreville and is rather sleepy. Being out late at night has a bad effect on him. I do not approve of it. I suppose you are living over old times in the little brick school house. Oh how I wish I could look into its pleasant room and see you all assembled there once more, but probably I shall see a great many things before I have the pleasure of beholding its dear old room again. But many were the happy days I spent beneath its humble roof. I have not time to write any more at present. It is cold here, and you must excuse my poor writing, my hand is very numb. So goodbye Jenny for this time, and believe me ever yours. James.

———————

Toward the end of November, the pace of camp life picked up dramatically. The final and long-overdue shipment of uniforms arrived, and on November 26 the men at last received their muskets. They were obsolete weapons purchased from European sources; James later wrote that the guns "would kick a man out of his hat every time he fired." Also on the twenty-sixth, an aid society, the Ladies of St. Joseph County, ceremoniously gifted the regiment a silken national flag fringed with gold tinsel, its staff topped with an eagle mounted on a bronze globe. "Never before," James exclaimed, "did the stars and stripes unfurled look more beautiful." Lastly, Brigadier General Don Carlos Buell informed Michigan on that same day that it was time at last for Colonel May to lead his men south.[41]

———————

Fabius, Michigan,

December 4, 1861

Ere this missive reaches you dear Jenny, I probably shall be on my way to Kentucky.[42] I came home last night on the evening train, and must return this morning. We are to leave tonight or tomorrow morning, so our officers told me. I was the only one who came. The captain told me he could not give me a furlough or pass, as he had no orders to let anyone go. But I was outside of the guard, and I have the privilege of staying out until the train came from Three Rivers tomorrow. I took the hint and came home. I should like to have seen you once more Jenny, but there is no chance for the present. But hope points to the bright future. I have no more time to write Jenny, as it will be car time by the time I get to Three Rivers, so goodbye Jenny for this time, from him who thinks of you often. James.

Camp Tilden,

White Pigeon,

Michigan, December 6, 1861

Knowing that you would like to hear from me once more, dear Jenny, before I leave White Pigeon, I avail myself of the present opportunity of writing you. The reason we did not go today, as I wrote you in my last, was because they could not get cars to take us away; so our leaving has been postponed until Monday, when we will bid adieu to White Pigeon for other climes. Since our last parting, you have been almost constantly in my mind. Dear Jenny, I often wonder where you are and what you are doing. If you are seated in the pleasant schoolroom busy with your books, or if your eyes do not occasionally wander to the seat I used to occupy, and then your thoughts leave you and wander away to other scenes besides those of the schoolroom. Jenny I have written imaginations long enough. But after all dear Jenny, it is sometimes sweet to imagine, do you not find it so? (But hope is the brightest of all.) There is nothing of much consequence transpiring in camp, only some of the boys are having quite a spree over payday. There are double guards stationed all around camp to keep the boys from running up town. But a great many get out and get intoxicated. Last night, Orderly Sturges and I went to the post office, and on returning we saw three of our boys go into a grocery.[43] We knew they were those who could not stand temptation, so we stopped and kept watch of them. They went from here into a billiard saloon, we followed them and finally prevailed on them to return to camp. One of them was badly intoxicated. He is one of the best boys in camp, ready to share his last penny with anyone who needs it, highly learned, capable of appearing in any society, but he has his failing and a fatal one it is too. His appetite controls him. Quite a sad accident transpired in camp about noon. We had just broke ranks from drill when I heard the report of a pistol in the direction of the sutler's. In a few moments, a soldier came running by me with a ball hole in his cheek, the blood flowing freely. I went over to the sutler's and he told me the boy had purchased a pistol of him and wished him to show him how to load it. It was one that loaded at the breach. He had placed the cartridge in the barrel, but had neglected to raise the hammer, and in returning the barrel to its place, the hammer came in contact with the cartridge, exploding it, the ball taking effect in the boy's cheek, passing down and lodging in his neck. The boy is now in the hospital. He is not dangerously wounded, he will probably get well.[44] Odd is writing a letter here by my side. Lem is having a scuffle with the boys. They have some tall times. It is getting pretty near time for roll call and I must close, so goodbye Jenny for this time, and believe me ever yours. James.

On the night of December 9, 1861, the 1,004 men of the 11th Michigan boarded rail cars and departed for Dixie, eager to fulfill their destiny in suppressing the rebellion. When the regiment prepared to cross the Ohio River, leaving the North behind, James King looked on as one of the captains, "drawing his sword . . . pointed to the Kentucky shore and intimated that Rebels over there were as plenty as blackberries in an old clearing. . . . No cartridges had yet been issued to us, and much discontent was felt after this speech." But the Michiganders received a cordial welcome upon their arrival in the upper South. The unit did not linger at Louisville—they would depart for Bardstown on the seventeenth—but the soldiers found time for some leisurely pursuits. James was among a number of soldiers who attended the theater and watched John Wilkes Booth perform on stage.[45]

Kentucky had struggled to remain neutral, yet found itself occupied by tens of thousands of soldiers from both sides. The task of saving the Bluegrass State for the Union fell on the shoulders of Brigadier General Don Carlos Buell, commanding the Department of the Ohio, and Major General Henry Halleck, spearheading the Department of the Missouri. Confederate general Albert Sidney Johnston barred their advance at Bowling Green. The subordinate brigadiers on both sides would determine the fate of the state: Union general Ulysses S. Grant at Cairo, Illinois, confronted Leonidas Polk at Columbus, Kentucky, while George H. Thomas opposed Confederate Felix Kirk Zollicoffer, who was positioned in advance of Cumberland Gap.

Louisville, Kentucky,
December 11, 1861

I am now in Kentucky, dear Jenny, in good health and spirits. We left White Pigeon Monday night at twelve o'clock for Kentucky. We took the Chicago road until we came to the Salem crossing. Here we took the New Albany and Salem road. We arrived here about daybreak. We changed cars, which took us some time. I enjoyed the trip very much. On our route, we passed some very large prairies. One we passed over in particular attracted my attention. It was very low, almost a swamp, not a tree or fence could we see. We passed through some nice towns, especially La Porte [and] Lafayette. The people all along the road flocked to see us and cheer us on. We arrived in Jeffersonville yesterday morning, just at the break of day. Here was one of the finest sights I ever saw. The sun was just rising in the east in splendor, the mighty Ohio went swelling on in power while on its broad bosom were four or five splendid steamers in sight, on the other shore were the Kentucky hills, and right opposite us was Louisville, a city of about seventy-five thousand inhabitants. Our regiment crossed the river about 9 o'clock in the morning, crossing in a ferry boat. We marched through the main streets of the city. Crowds of people thronged the streets, ladies with their silks and satins, ribbons flying, men in military dress, Negroes by the wholesale. I saw more colored people yesterday than I ever saw in my life, and most of them slaves. There are some massive and splendid buildings here. They gave us a good welcome. They thought we were the best regiment that had passed through. Troops are continually rushing forward. There is nearly one hundred thousand Federal troops in this state. We move tomorrow to Bardstown, about seventy miles from here, where we expect to see secesh if they do not leave before we get there. I have a very poor place to write, so I must close. Goodbye Jenny, and remember me as ever yours. James.

Louisville, Kentucky,
December 13, 1861

The dusky shades of night have settled over the Earth's broad bosom, and I am seated in the tent by a camp chest, writing to you. I am enjoying myself hugely, to use a favorite expression. The weather here is almost as mild as summer. The people here tell me that they have only had a few inches of snow, and that went off the next day after it came. Yesterday I got leave of absence and strolled around the city. I passed through several of the streets. I saw many splendid buildings. Among them were some private dwellings which were very beautiful. But the most splendid of all was the city post office. It is built of hewn limestone, and [is] a splendid pile. It is situated near the river. The streets are all neatly paved and dry and hard. But their dooryards I admired most of all. They are raised 2 or three feet above the sidewalk. A limestone walk leads up to the mansion steps of limestone to ascend from the streets. The grass is as green and fresh as in the month of June, and are surrounded by an iron fence resting on a limestone foundation. Standing in one of the main streets, I could see persons of all classes and professions. Here comes some ladies in their carriage who are counted among the upper tens, with a Negro driver perched on a box whose business it is to take the ladies wherever they wish to go. You can tell the business man by business airs he puts on. There goes the dandy with his cigar, his hands in his pockets. Here you see the Negro with his mule team of four or six mules, riding one and driving the others. There is a great many colored people here. They seem to be well dressed and happy, but I heard one of them say they would all run away if they could get a chance. I wish you could be here & see all I have seen since I have been here. Then you would know more than I can describe. But Odd has written you once or twice, and perhaps he will make up what I lack. Troops are constantly arriving. There will be something done before long that will put a damper on secesh. The first charge we made after getting in secesh was on a haystack. We needed something to put in our tents for bedding. The colonel told us to take it, and if he was a Union man, he would pay him for it, if not, he would not. We enjoy ourselves better here than in Pigeon, have things more like men. General Buell has his headquarters here in this place, and the Rebel general Buckner used to live here.[46] He was very wealthy, but his property is now confiscated. There is only one thing that makes me feel sad, and that dear Jenny is to be separated from those I love as dear as life. I hope you and Rebecca still continue to like the school.[47] I expect that you must have some great times. I have no more time to spend writing now, so goodnight dear Jenny, and remember me as ever yours. James.

Camp May,
Bardstown, Kentucky,
December 22, 1861

As the thoughts of the past come crowding on my mind, I remember that I have friends in my dear home in the North who would like to hear from me and know that we are still in the land of the living, as I had heard dear Jenny that the news had reached St. Jo that we were all killed and taken prisoners, but we have had no fight yet or hardly seen a secesh. We left our camp at Louisville Tuesday morning and marched on foot to this place. On our route we passed some very fine country, besides some very rough. The country was level for about fifteen miles after leaving Louisville, and under a high state of cultivation. It would compare very well with the best land of the North. The plantation houses resemble the farmhouses of Mich., only more splendid and of larger style. After this, it began to grow hilly and rocky, the roughest I ever saw. The roads here are not like ours, they follow the ridges. They go more for the levelness of the thing. The second morning after we started from our camping ground, we traveled about 3 miles when I am sure we were not more than 1 mile from where we started. Our present encampment is situated on a grassy slope, at the foot of which flows a babbling brook. A spring which gushes from between two limestone rocks furnishes us with water. There is no wells here, but springs every few rods. The people all along the road came to meet us and bid us Godspeed. This did not look much like secesh. When we entered the village of Bardstown, the people thronged the streets, cheering us enthusiastically.[48] There is no enemy within thirty miles of here. We will probably stay here about a week, when we will be pushed on farther down in Dixie. The weather has been very fine until today, when it seems as though the very heavens had sprung a leak. The rain pours down in torrents and everything is all afloat. This makes it dreary to be penned up here with 18 or twenty men in a small tent. Our rations are army crackers and coffee, sometimes fresh beef. Bacon is the principal meat.[49] You who are at home surrounded by every comfort can little imagine the scenes of a camp life. All of the boys from near home are well and in good spirits. Odd lies by my side telling the boys some of his trials. I think often of home and the good times you must have this winter. I get to thinking of this and forget myself, and think 'tis hard that I should be so far from you. Then duty stares me in the face and tells me to press onward.

> Not few are the burdens of life
> Then load it not with heaviness of spirit

Dear Jenny sometimes my heart gets heavy and all seems dark.

> But hope can pierce with quickening ray
> And all those depths are lighted[50]

I have no more to write now, so goodbye dear Jenny, and remember me ever yours. James.
Excuse poor writing, as I have written this on my knee. Give my best respects to your father and mother.

Camp May,
Bardstown, Kentucky,
December 26, 1861

Dear Jenny, yesterday was merry Christmas. Even in camp, 'twas not forgotten.[51] You would hear in every tent, I wish you merry Christmas. In the forenoon we drilled in the manual of arms. We went about eighty rods from camp in a piece of beech woodland, where we were practiced shooting at a mark. The day was like summer. Oh such delightful weather for winter I never saw. We have only had one day that was any ways uncomfortable, and that was a rainy one. There has been no snow here since we arrived, but the people here say this is uncommon fine weather for this time of year. At dinnertime we went to camp, where we found dinner awaiting us. Odd has been chosen as one of the cooks for our company.[52] There is three of them. I wish you could see him fly around, he acts like a man of business. After partaking of our dinner, which consisted of boiled beef, coffee, and hard bread, I obtained a pass from the colonel for Orderly Sergeant Sturges and myself to go up town. After passing the guard, we took the road that led to the village. Every few rods we would meet a Negro. Some were riding mules, others mounted on ponies, some were on foot, all as happy as could be, for this is the Negroes' holidays. They are all dressed in their best, many of them splendidly. They have nice shirt bosoms, starched collars. Their boots are blacked, and a white handkerchief is almost sure to be seen protruding from some one of their pockets. They go from one plantation to another thus, gathering and feasting and dancing the time away. They do no work from Christmas Eve until the day after New Year's. Every little child we met in the street would shout, Christmas gift. They do not wish you a merry Christmas as we at the North do, but it is all Christmas gift. The streets were full of planters and country people purchasing gifts for the holidays. After passing through the principal street of the town, we made up our minds to visit a cave which lay back of the town. We passed some large rocks that hung frowning over our path. The entrance to the cave was a narrow passage of about 3 feet square, through which we crawled about 20 feet when, having no light with us, we made up our mind to go no farther. We faced to the rear and soon found ourselves in daylight again. The people say it is several miles long, extending up under the village. Some of our boys prepared lights and went in it about half a mile under the village. They found a stream of water which flowed through here. It was cold and damp, making some of them sick. They said it had a very disagreeable smell, and they were glad to get out. I thought of home often yesterday and wondered how my friends were enjoying themselves at home. Such days will remind one of home. I could see them as they moved around in each well known room. The bright fire glowing in the stove. I saw the evening meal prepared, and then they drew around the festal board. They talked of him whose place was vacant. After this, the table was cleared and everything in its proper place, the candles lighted, the stand moved from its accustomed place in the corner, the lights placed on it, then each took their places around it. What a glorious thing dear Jenny is imagination. It is at least with me, since I enlisted. I am enjoying camp life better than I expected in secesh. We have plenty to eat, such as it is—hard bread, beef, and coffee. I can hardly tell how it would seem to sit down to a meal in a civilized land. When our meals are ready, each squad of men takes their share (a squad consists of about eighteen men) into their tent, and all pitch in. Every fellow for himself. Our dishes consist of a tin plate, a knife & fork, and a tin cup. After the meal, our dishes are washed by the cooks and put into a large chest called a mess chest, where they remain until we wish to use them again. Our boys are in good spirits most of the time, only when it comes wash days, when they are a sorry looking set. I don't blame the women for being cross on wash days now, since my experience in that line of business, if they have to labor under as many disadvantages as I do. I wish you could have seen Lem the other day doing his washing, he used some tall exclamations. I believe if Lem was home again he would stay there. Odd is perfectly at home. I have not time to write any more at present and I am ashamed of what I have, but you must remember, letter writing in camp is rather inconvenient. The paper I have written on is a sheet that I have carried in my knapsack a long time. I could not get up town to get any more. But I thought, dear Jenny, you would like to hear from me [even] if it was not on a very clean sheet of paper. I have waited and watched for a letter from you, but have been disappointed. I have written 15 or twenty letters since I left White Pigeon, but have not received a single letter. I hope you will write often Jenny, but I must bid you goodbye for this time. Remember me as ever yours. James.

> Camp,
> four miles from Bardstown, Kentucky,
> December 29, 1861

Dear Jenny,

I received your letter in due time, and you cannot imagine the joy it brought with it. I had looked anxiously for a letter from some of my friends, but was disappointed until night before last, when I received yours, one from your father, and one from my brother [John] in Waukegan.[53] I have had none from home yet. I cannot imagine why they do not write. We left our encampment at Bardstown yesterday morning and moved four miles farther into secesh, where we have taken up our position.[54] How long we will stay here, I cannot tell. It is a very pleasant situation where we are encamped. To the north of us is a fine piece of woods covered with oak, hickory, and maple trees. A piece of woods also bounds it on the west. To the south in front of us is a large, level field. Far beyond this, in the distance, you can see the tops of mountains which rise above the hills. You can tell them by their cones rising far above the woods, & covering their tops are tall evergreens, which makes a very magnificent view. It was all new to me, I never saw anything like it before. To the east of us is the turnpike. Across this is a cleared field, and on this is encamped four regiments of infantry. Three of them are from Indiana, and one from Ohio. I wish you could see this camp at night when their lights are all burning. It is a splendid sight. I have been writing all the forenoon for Capt. Oakes, and while he was eating his dinner and using his table, I took this opportunity of writing you this letter. We have things very comfortable here. We have stoves in our tents, which makes it seem [a] little like a living. Goodbye dear Jenny. From him who will never forget you. James.

1861 drew to a peaceful close, with the regiment enjoying tolerably good health. 1862 would be a very different year, commencing with the onset and raging spread of terminal diseases and ending with the deadliest day of combat the 11th Michigan would ever experience.

CHAPTER 2

Rations and Coffins

January–April 1862

❦

*Rather would I fall on a battlefield than
end my life in a hospital.*

The enthusiastic recruits from southwestern Michigan were spoiling for a fight, but their first battle pitted them against a ruthless enemy that offered no path to glory: disease. Ignorance and negligence of the need for sanitary living conditions were rampant in Civil War armies, particularly early in the war, and triggered disastrous consequences. Units that hailed from lightly populated areas often suffered epidemics of diseases such as measles and smallpox early on—sometimes just after the soldiers deployed. Such was the miserable case of the 11th Michigan. James King would remember the long weeks at Bardstown as "one of the darkest spots in our soldier life," grimly noting that "we received rations and coffins in the same wagon-load from the village." Before they even met the enemy, the Wolverines faced death on a frightful scale.[1] Just days after settling down at Camp Stoughton, the ailing regiment would be ordered into isolation at Camp Morton—a name that would soon become synonymous with disease and suffering.

———

Camp Stoughton,
four miles south of Bardstown, Kentucky,
January 2, 1862

Dearest Jenny, 'tis nearly midnight and I am seated in my captain's tent watching by the cot of my second lieutenant, who is very sick with the measles.[2] My thoughts wander tonight. I cannot keep them with me. All without is dark

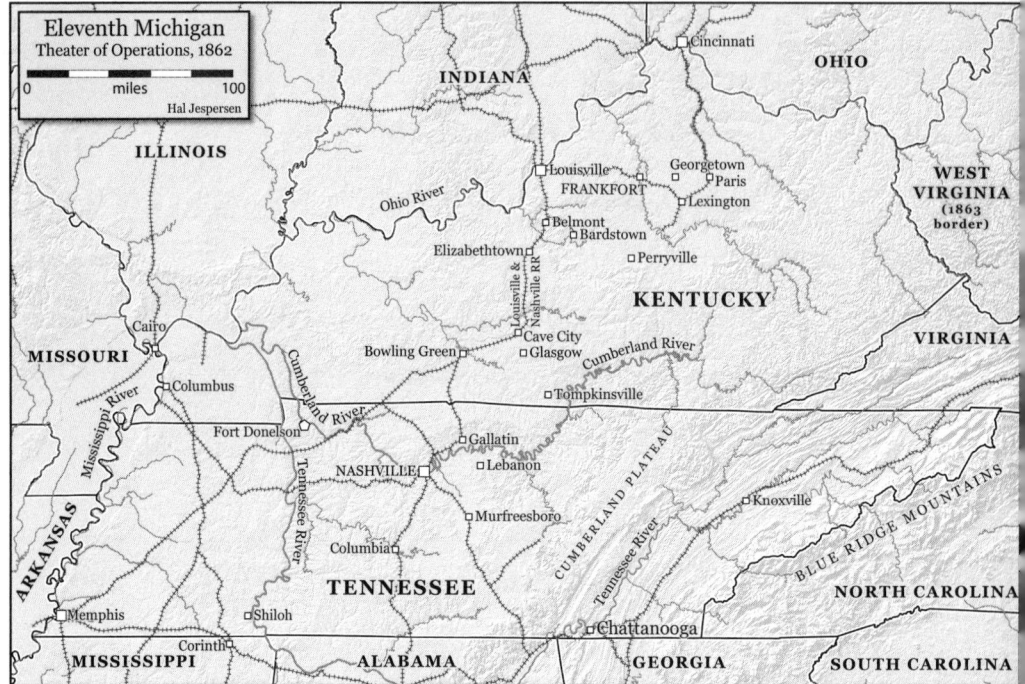

and dreary. The rain falls in torrents. The winds have a pitiful moan. 'Tis such a night as will set one to thinking, if they ever had a thought in all their lives. My mind has wandered back to my early life, when I sported, a free and careless child. When all was sunshine, not even a passing cloud marred my childish joys. Those were halcyon days, never to be forgotten. But dear Jenny, the happiest day I ever saw was when you told me you loved me. Never shall I forget that.

Your likeness now lies before me. Dear Jenny you cannot imagine the pleasure it has given to possess a thing so dear. Could I but see the original and clasp you in my arms as I did at our last parting. But dear Jenny it will be many long months, aye; perhaps years before that fond hope can be realized. But as long, dear Jenny, as there is a traitorous hand uplifted, I shall be found at my post. Oh, will it not be sweet, when war's alarms are o'er, to return to home and friends. You said my letters were doubly dear to you. Just imagine Jenny how dear a missive is from you. There you are at home surrounded by friends. Here I am in the crowded tent without friends. I said without friends, no that is not so Jenny, all are my friends, [but] after all they are not the friends at home. I was pleased to hear that you were having fine times and enjoying your school. Also, that you and sister had attended the school at Three Rivers with uncle.[3] I know you must have had a good time. I think after you get acquainted with him, you will like him very much. I suppose you are not quite as backward about going to father as you used to be. I hope you go often. I know you are a welcome visitor. It is very pleasant country round our camp, and no danger from secessionists. I am now acting as company clerk again, Holbrook having received the appointment of brigadier general's clerk. I cannot write any more at present, so goodbye dear Jenny and remember me as ever yours. James. Let no one else see this as I intended it for only you—the rest of the boys are well.

Camp Morton,

near Bardstown, Kentucky,

January 12, 1862

Dear Jenny,

'Tis Sunday and I am seated in the capt.'s tent penning this epistle to you. I have written 2 or three times since I received any answer from you, and begin to wonder why you do not write. I am sure you have written before this time. Probably I write too often; though 'tis the same with those at home. I have had only one from home when I have written several. It is very discouraging to us soldiers to write five or six letters a week and only receive 1, or perhaps not any, in return. It cannot be that I am forgotten; no, I cannot think that. I know dear Jenny that a soldier is considered by some below the ordinary rate of men but I think not, when men are engaged in such a cause as we. I do not think you think so.

> No, Jenny, dear my life must be,
> Since it is worthy care from thee;
> Yet life I hold but idle breath,
> When love or honor's weighed with death[4]

We are encamped about two and one half miles northeast of Bardstown on a farm known as the Bowman place; but is now owned by a man by the name of Jackson, said to be a brother of Jackson, the man who killed the noble Ellsworth. Jackson is now under arrest for shooting a man.[5] When we arrived here, we found everything deserted. A large plantation house stood on the premises, surrounded by numerous outbuildings. There has a great display of taste been laid out on this farm. A fine lawn extends in front of the house, and set with a fine grove of locust trees. Capt. and I took a stroll last night through this lawn and some of the adjacent grounds. It was one of these lovely nights when

> The moon like a rick of fire
> had risen o'er the dale[6]

But there was a charm in this I never felt before. It seemed to me I never saw the moon cast so mellow a light over the Earth. There lay the wide, stretching lawn covered with green turf, dotted here and there with the dark and leafless locust. The mansion, painted white, glittered in the moonbeams. To the east of this burned the campfires of our regiment. On the north lay a stretch of heavy timber, which looked in the pale moonbeams like a dark and frowning wall. I see many things here that is truly romantic. Many kinds of timber which I had heard of often, but never saw before, are here. Among the several kinds is the famous gum tree of which we often sing. The persimmon & hackberry you also find here. The mistletoe is also here. It took me by surprise. I thought it must be a small tree by what I had read, but I find it a bough, that is all there is of it. It never touches the ground. It grows upon the tops of trees, but generally on walnut or ash. You can see all through the tall and leafless woods their evergreen plumes. They tell me where they are thick, they often kill the tree. I wish I could describe some of the scenes I have beheld here. There is no hills here as in Mich., but ridges and ravines. You will find in traveling broad belts of level land, well cultivated. Then you will come to a dark and cheerless ravine, walled in on both sides with huge rocks so steep in some places, you would need a ladder to ascend. Looks dismal enough. In the bottom of these generally runs a whimpering brook.

The climate is very mild, only a few flakes of snow have fell here to remind us of winter. But we have a great many dismal rains. The soil is of heavy clay, and so much rain makes it very disagreeable getting around. I thought I had seen mud, but I never saw any to compare with this. Our tents are a very good protection against rain. We have a bakery established here which makes it very handy for us, it goes far better than mixing our own bread. A couple of the boys do our washing for us for three cents apiece. This was the greatest trials we had, baking & washing. You may think these were light trials, but deliver me from anything of the kind. Especially in camp where there is no conveniences for it. Alden is still in the hospital, but is getting better. Lem is well. I have time to write no more at present, so goodbye dear Jenny, and believe me ever yours. James. We shall probably stay here quite a while. I think the Union cause is growing here.

Camp Morton,
near Bardstown, Kentucky,
January 17, 1862

Dear Jenny,

I received your letter of January 6th last night. It came to me like a sweet dream on a stormy night. One, to appreciate a letter, must be separated from friends and home. They are welcome messengers to me always. I hope dear Jenny you will write every opportunity you get. We are still in Camp Morton near Bardstown, where we will probably stay until the measles abates. This disease rages considerable in our regiment.[7] Odd has been very sick with them but is most well now.

The smallpox has nearly run out. There being only 3 or four cases, and they will be brought to camp as soon as new clothes can be provided. I expect you hear strange stories about us, captain just received a letter stating he was dead, and both of the lieuts., when in fact he has not been sick a day. Other letters state that we are dying off thirty or forty a day, when in fact there has been very few deaths. I cannot see where such reports originate. There is a general forward movement of all the troops in the Federal command, and you may expect to hear soon of a mighty blow being struck. We received news yesterday morning of the defeat of Humphrey Marshall by Col. Garfield.[8] I will not give you the particulars of the battle, as you will receive it in the papers ere this reaches you. The duties we have to perform each day is as follows. At daylight the reveille is sounded, when we get up, pack our knapsacks, and attend roll call. The company is formed by one of the sergeants and each man answers in person in his own name. When this is through with, an inspection of arms and knapsacks takes place. Then we break ranks and prepare for breakfast. At nine o'clock knapsack drill takes place, which lasts until nearly noon. In the afternoon we have battalion drill, which is superseded by the colonel. At four o'clock dress parade takes place, when we are disbanded and return to quarters. At 9 o'clock in the evening another roll call is made, when we are free from the duties of the day.

Lem is well, he drives the company team. I have no chance of writing any more at present, so goodbye from your James.
Give my best respects to your father & mother.

Camp Morton,
near Bardstown, Kentucky,
January 21, 1862

Dear Jenny,

I received your letter last evening, and you cannot imagine the joy it brought. I also received one from sister. I did not think, dear Jenny, you had forgotten me, but it did seem strange why you did not write. But of late I have no reason to complain. The weather still continues warm. Though yesterday morning, upon rising from our beds and looking from the door of our tents, we saw the ground covered with snow. This looked like home. But we did not behold this sight long.

Before evening it had nearly disappeared, and instead of the beautiful mantle of white which we beheld in the morning covering the bosom of the Earth, this vile Kentucky mud met our gaze. Oh Jenny you know nothing of mud. The ground around our camp got so bad, we tore the fences down and made roads of them. This is quite an improvement, only think railroads all around our camp. You may think that our camp must be woeful looking by my speaking of mud, but you would think different if you would see it. Each company's tents are set on a street by itself. These streets are all set with evergreen boughs, which gives it a fine appearance. You wished me to tell you if I have to perform guard duties. I have only stood guard four times since I enlisted, and am also excused from a great many other duties. I have been very busy for a few days making descriptive lists of the boys in hospital, and some are being discharged on account of physical inability to perform their duties. This makes a great amount of writing. Alden is gaining a little, he is so he can walk but is pretty sick yet. If he does not get any drawbacks, I think he will get along. I go and see him every opportunity I get. There has several died since I last wrote. Jenny 'tis a horrible sight to visit the hospitals and see the poor fellows languishing in beds of pain.[9] Sickness will kill more than the leaden ball. There is ten or twelve lie buried in the cold damp ground near our present encampment. They enlisted [as] full of life and hope as any of us, but alas, they have fallen before they met the foe. Death is horrible, let it come in any shape it may, but dear Jenny rather would I fall on the battlefield than end my life in a hospital. I still continue to be healthy. I never enjoyed so good health as I have had on the tented field. You must excuse poor writing as I had to hurry in order to have it go out tonight. Oh, how I should like to be with you and enjoy some of the good times you are having. Write soon. Goodbye. From your James.

Camp Morton,
near Bardstown, Kentucky,
January 25, 1862

The morning sun shines warm and pleasant. The breeze sweeps lightly by. Bluebirds begin to sing and everything seems like spring, and here it is in the dead of winter. I suppose you at home are having good sleighing now. A few of our boys who were left in White Pigeon on account of sickness arrived in camp last night. They report good sleighing and pretty cold weather. How strange it seems. The rains here are very disagreeable, but when we do have a sunny day 'tis indeed beautiful. The sun seems to shine brighter here than at the

North. Yes dear Jenny there are many attractions in this life, but quickly would I exchange my situation here if I did not consider it a duty I owe my country. I am in hopes that this war is of short duration. The Rebels are losing ground every day. I think 'tis plain secession has seen its palmiest days. The loss of Zollicoffer has changed the spirits of their dreams. They had confidence in him as a leader. But he has gone to his long home.[10] Our reg. is still encamped near Bardstown. There is no other regiments near us, all the rest having been sent down to Green River where active operations are now going on. 'Tis hard to be kept here idle when our brethren are in the field, but I suppose it is all for the best. This delay I cannot account for, [unless] only [for] sickness. The smallpox has nearly left our camp, but the measles are raging considerably. Alden is yet sick. He was taken first with the measles and had almost recovered when he was taken with a fever. He also received a letter from his brother Lewis giving an account of his father's death, and I think this had a tendency to make him worse.[11] He is a pretty sick boy, but I think he will recover. All of the rest of the boys are well. I received your welcome letter last night. I wish I had time to write more, but have not, write soon and believe me ever yours. James.

> Camp Morton,
> near Bardstown, Kentucky,
> February 4, 1862

I received your letter this morning among four others, and I can truly say dear Jenny it gave me more pleasure than all the rest together. Dear Jenny I did not mean to write anything that would mar your feelings; if I have, I am very sorry. It pained me to think that you thought, for a moment, that I thought you had changed. No Jenny, not for one moment has it ever entered my mind. Jenny never let it enter your thoughts again. If I had not thought you were true, I should never have loved you. I got a pass with several of my comrades to take a stroll this afternoon, among them was friend Wells. The day was very fine, one of those that makes you feel in love with everything. We strolled from camp in an easterly direction. We first came to a beautiful wood of large whitewood trees; through this we passed and came to a small creek, which ran sparkling and flashing over a bed of solid rock. I wish I could describe to you this little stream as I saw it this afternoon. I never saw anything so beautiful. Sometimes in following its windings we would come to a place where it went rushing over the rocks eight or ten feet high, then it would go dreaming on as quietly and gently as though its course was almost smooth.

Hill, nor rock, nor stone,
Laid in the path to me unknown.
But a forest land, which varying still
With ridge, ravine, rock dale and hill[12]

In the balmy air of the afternoon we wandered along this beautiful stream, then over undulating swells that swept down and away again ere they rose to the prominence of hills. Then we would come to a large landscape stretching away as far as the eye could reach, and bordering on this, the dense forest would rise. We wandered about this way until nearly sundown and then returned to camp. Alden is getting a little better; Lem is well. I wish I had time to write you a good letter, but I have no chance. It is getting late, and I must close. The boys have all gone to bed, and I am in the capt.'s tent. Christopher Haight, our first lieut. who was married soon after we enlisted, is not expected to live.[13] You said Sunday you wondered if I were thinking of you. Dear Jenny there is not many hours pass but what I do. I consider your happiness is dear to me as life. Never think as you did again dear Jenny, and remember your James.

Camp Morton,
near Bardstown, Kentucky,
February 10, 1862

Dear Jenny,

While seated in my tent, surrounded by my comrades, listening to their talk and watching their actions, my thoughts turned to you, and I thought I would spend the evening writing you a letter. If I could only see you, I could tell you more in one hour than I can write in a week. But the pleasure of seeing you I do not expect will come right away. I live in hope. Though I do not think but what we will be discharged in a year at the longest, and maybe sooner. I got a pass yesterday to ramble in the woods. I took a southeasterly course. I had went about one mile from camp when I came to a rude plantation house surrounded by a row of Negro huts.[14] I took a path which led back of the house down to a ledge of rocks. I soon found myself at the entrance of a cave. There was a large room at the entrance, about 25 or 30 feet square. A narrow entrance led from this room. I followed this several rods, but it being dark, my lantern not giving sufficient light, I concluded I would go no farther. A stream of pure water issued from this cave. I never had such feelings before in all my life as I did when I stood in that cavern of rocks. It was not fear, but if I felt my weakness, it was then. I wish you could have been with me. Enclosed you will find a small bough of the far-famed mistletoe, which I plucked in my rambles yesterday. The drum is beating for lights out and I must close, so goodbye. From your James.

––––––––––

Christopher Haight succumbed to typhoid, triggering a flurry of promotions in Company A. James King was granted the rank of corporal. He made no mention to Jenny of his promotion. By mid-February the regiment had lost more than fifty men to disease. Eleven Michiganders perished between the ninth and seventeenth of February alone.[15] On February 15, for the first time, the war claimed someone close to James and Jenny.

––––––––––

Camp Morton,
near Bardstown, Kentucky,
February 16, 1862

Dearest Jenny,

It is with sadness that I undertake the penning of this letter to you. Your cousin Alden is no more. He died yesterday morning at half past eleven. The last words he spoke to me were day before yesterday. I went to his room in the morning, he looked very bad, but said he felt better. He talked and said, if there should anything happen to him, I would let his folks know it. I visited him again yesterday morning and found him so far gone as to be unable to speak. But he seemed to recognize me. Poor fellow, he suffered everything. He was taken first with the measles, recovered so that he was able to be out. The typhoid fever then set in. He was sick with this a long time when the erysipelas broke out on his face, which carried him away.[16] We dug his grave beneath the sheltering boughs of an apple tree and laid him in his narrow bed to rest. There are about forty of our regiment laid in the cold, damp ground in this orchard. It is a painful duty to lay a comrade to rest, but one of which we are often called on to perform. Little did I think last summer that I should be called on to lay him in his grave so far from home and friend. 'Tis hard, but we cannot tell today what the morrow will bring forth. The rest of the boys are all well. Lem feels very bad. We are still encamped near Bardstown and there is no prospect of our leaving right away. I think the general health of the regiment is improving. May the good angels keep you, dear Jenny, and remember me thine. James.

Camp Morton,
near Bardstown, Kentucky,
February 19, 1862

It has rained nearly all day, a continuous shower. I have listened to the patter of the soft rain on the—I came near saying roof, well, I have listened to the patter of the soft rain on cotton cloth. Picture to yourself dear Jenny one of these cotton habitations. A stove in the center raised up from the ground about 6 in., and a pipe projecting upward to carry out the smoke. The ground [is] covered with five or six inches of straw, which serves for a bed. Around the edge of the tent is a row of knapsacks and blankets. Then picture to yourself the inmates, say about 12 or fifteen U.S. soldiers, variously occupied.[17] Some writing, others reading, some mending, a few singing, and some talking of home. I believe there is more said about home than anything else. I never say much about home, but that is not to say I do not think of it often. If you can imagine all this to yourself, you will have a pretty good idea of our home. I received your letter Monday, & one from sister also. You cannot imagine the joy they brought me. Last Sunday evening I went to a farmhouse to visit our lieut., who is sick. Their names are Unsils and are what are styled abolitionists.[18] They are good Union people, doing many things for the comfort of our soldiers. A comrade by the name of [Daniel] Rose accompanied me.[19] We arrived just as the family were sitting down to supper. A bright fire blazed in the fireplace. The table was spread with the good things of this Earth. I tell you Jenny it made me think of home. The people shook hands with us, bade us welcome, and invited us to supper. We did not need much urging. The family consisted of Mr. and Mrs. Unsils, his two sons, and daughter in law. After supper we went into the next room, and found Lieut. Fisher and Doctor Wheeling [Arvin F. Whelan] both sick but in very good spirits.[20] They were very glad to see us. Lieut. told me he had first rate care, only they wanted him to eat all the while. We stayed with them until nearly 9 o'clock, when we returned to camp. Now we were in a fix. The countersign had been given out before this time and what were we to do? I knew if I could find one of our boys on guard, it would be all right. So I marched up to the guard line followed by Dan. The sentinel saw us and challenged us with who comes there. I answered, a friend. He replied, alright Jim, rather late to be out, but pass in and keep cool. He knew my voice. The more I see of Southern life, the less I think of it. There is a lack of enterprise and industry. The whites are a puny, pale, sickly looking set, acting as though life was a burden. There are no schools here, and I have been told that but very few of the inhabitants can read and write.[21] But the country is very beautiful, and for natural scenery cannot be surpassed. You warned me not to run recklessly into danger. I remember the promise I made to you Jenny, but when I have friends who are sick and wish to see me, I consider it is my duty to go. Only think Jenny if it was me instead of them. I used to go and see Alden often. He was as good a friend as I could wish to have. He received a letter from Letitia the day before he died, and requested me to answer it.[22] I done so. I also wrote his sister Ruth of his death. The rain has ceased falling and [it] is pleasant once more. Bluebirds are singing, the grass begins to start, but before tomorrow morning it may snow. One day here, it was so warm Capt. Oakes was obliged to eat his dinner outdoors in the shade. The next morning, the ground was covered with snow. But the snow never stays more than one day. We have not had a day now we could call cold, only disagreeable. I must close my letter or lose my supper. With my best regards for your father and mother, I remain ever yours. James.

Camp Morton,
near Bardstown, Kentucky,
February 23, 1862

Dearest Jenny,

It is the recollections of the happy past that causes me to attempt the penning of this missive to you. I am in no mood for writing today, and but for your own dear self I should not make the attempt. As it is, you must take such a letter as I can write. Though I have often heard it remarked, the more sensible a man, the duller his letter, & I think Jenny if you take this letter for its dullness, I must be pretty sensible, though I am afraid that this doctrine will not always hold together. It is Sabbath's holy day, and how different the scenes that are passing with you to what they are in camp. It is now about 10 o'clock, your morning work is probably done and you are enjoying the quiet scenes of home. But here, all is noise and bustle. Preparations are busily going for Sunday inspection. Each Sunday there is an inspection of all the soldiers, arms, accoutrements, & clothing. I do not think that there are half of the boys that think of the Sabbath.[23] I am afraid Jenny this war is having a bad effect on a great many, some are forming habits which I am afraid will be lasting. It is not only the men, but officers. There is hardly a man but what use intoxicating spirits when they can get it. They say when the war is over, I will quit it. But appetites formed here are just as lasting as those formed anywhere. I think the best plan is touch not, taste not, handle not. It is the moderate drinkers that make the drunkards. It is true dear Jenny there are many vices in camp life, and there are many things which will tend to improve one if lightly applied. I think it will not be many months before we will be discharged. The war is assuming a very important feature. News reached us last night of the capture of Floyd

Arvin F. Whelan.
(Courtesy Archives of Michigan)

and three thousand prisoners. Governor Harris of Tennessee has ordered the people of that state to lay down their arms.[24] If this is so, traitorism is nearly at an end. One of my companions has just come to have me take a stroll with him. It is very pleasant without, and I guess I will accept the invitation. I should like very much to have you here and enjoy some of the beautiful scenery. The grass is getting quite green in some places. Jenny I believe if my friends were here, I would choose Kentucky for a home. Not that I love it better than Mich., but the climate is what makes the difference. I received your letter yesterday, and one from sister. You wished me to write a piece for your paper. I should like very much to do so. But this is a poor place to write. There are a great many things I would like to talk with you about, but I must postpone for this time. With my kindest regards for your father and mother, I remain your James.

Camp Morton,
near Bardstown, Kentucky,
February 28, 1862

Dear Jenny,

The birds are caroling their sweet notes to us. The sounds of their welcome strains is delightful to hear. Since I last wrote, we have moved our camp one mile nearer town. It is one of the most pleasant places I ever saw. A large brick house stands to the rear of our camp. In front of this house is a wide lawn covered with shade trees, mostly hickory and walnut. Most of these have their tops covered with the green boughs of the mistletoe. On this lawn we pitched our tents. In front of the camp is a ravine through which runs a sparkling brook. The sides of this ravine is piled up with shaggy rocks. Across this ravine, about the distance of thirty or forty rods, stands the splendid mansion of Ex-Governor Wickliffe.[25] This is the most tasty residence I have seen since we left Louisville. The house is large, two stories and a half high, and built of brick. The grounds around the house are set with evergreens and trees of every description. I have seen many splendid groves, but none that will compare with this. I visited the grounds yesterday but did not see any of the inmates, only a couple of Negroes who were plowing in the field adjoining the house. Governor Wickliffe is now in Washington & is reported to have two sons in the Rebel army. There is no news in camp, only that there is some prospects of our moving before long, but in what direction, I have no idea. I often wander away from the busy scenes of camp, and then dear Jenny I ponder over bygones. I often find myself thinking of the happy times we have passed together. I often wonder,

When again I'll stand with your lovely hand clasped warmly in my own
And hear your fond and much loved voice
In its low and quiet tone

I look forward to the future with a hopeful heart. No more at present, goodbye. From your James.

———

Following the surrender of Forts Henry and Donelson to Ulysses S. Grant, Union forces forged ahead into Tennessee. In early March the 11th Michigan was ordered to guard a vital supply line, the Louisville and Nashville Railroad. "There were no regrets," James declared, "when we exchanged this camp of sorrow and death."[26] Colonel May's men were scattered across intermittent points guarding the railroad while the bulk of the army embarked on one of the war's bloodiest campaigns.

———

Camp Morton,
near Bardstown, Kentucky,
March 6, 1862

Dear Jenny,

It seems more like winter for the past few days than any we have had. I think after this storm is over it will be warm and pleasant again. But Jenny when I see the mantle of white which covers the Earth, it brings old scenes to my mind. Our regiment moved this morning for Elizabethtown, a distance of twenty-six miles. I will remain here probably five or six days, when I will go on. I am acting as hospital orderly. There are about two hundred who are unable to carry knapsacks. They will wait until things are regulated in camp and will then be sent on by railroad. The position I occupy is a good one. All I have to do is write, which keeps me busy about three hours in a day. I have a good warm room to work in, a bed to sleep in, and everything comfortable. The one who occupied the place was taken sick, and I will act until he gets well. I pity the boys who will have to camp on the snowy ground tonight, a great many will take colds and sickness will follow. If it was not for sickness, camp life would not be so very bad. But a great many more die of disease than are killed in battle. You who are at home surrounded by every comfort can have no idea of the suffering that many poor soldiers have to undergo. I could paint a picture Jenny which would cause tears to flow, but I will not. I have letters from friends very often who tell me about the good times they are having attending festivals and balls. It does not seem right that a brother or sister should be reveling in a festal ball when perhaps the noblest of their family may be suffering in the crowded hospital or gasping his last breath on the bloody field of battle, sacrificing his life blood to sustain the rights and liberties of a free government for them to enjoy in after years. I do not believe Jenny in putting on garments of mourning because we are fighting for our liberties. But I do think all necessary expenses ought to be avoided. This war costs money as well as lives, and it is from the people that this must come. A national debt is being incurred which will take years of toil to remove. That national ball given by Mrs. Lincoln was a wrong thing.[27] I have a paper illustrated which gives a minute account of the occasion, and the splendor and magnificence displayed. If they would have paused and considered how many poor & suffering soldiers were at that very moment sick and dying for the comfort which they were so foolishly wasting, their jollity would have been turned to sadness. Ah Jenny if they could see the groaning hospitals eject their dead as I have, it would cause their limbs to chill and veins run cold. I did not intend writing such a letter as I have when I commenced, Jenny, but my thoughts were turned in this channel and I have written more than I should. 'Tis getting [to be] mail time, and I must close in order to get this off. Goodbye Jenny, & remember me as ever yours. James.

Tell your mother I should like to receive a letter from her.

Bardstown, Kentucky,
March 11, 1862

Dear Jenny,

I have been very busy writing, making reports for hospital, writing papers, and suchlike. I received an order from the col. today to report myself to headquarters. I will go tomorrow by railroad to Belmont, where two of the companies are stationed. The col. has his headquarters there. My company has gone to Elizabethtown, about 12 miles from Belmont. What he wants of me I do not know, but I think to act as clerk in the hospital there. I received a letter from sister last night, she wrote me a good long one & among other things she wrote, said she supposed it would not interest me unless she put in the word Sarah. Well I confess, there is much to interest me in that word, but 'tis not the word alone, 'tis the one who bears the name. Dear Jenny I hope the day is not far distant when we shall meet. But the longer the separation, the stronger the tie that binds us to each other.[28] The sun may rise and set many times, the seasons pass in their annual rounds, but never will I forget the one I promised. I would be glad to be among my friends Jenny, but we must submit to many trials which it is painful to bear. But as you say, there never was a cloud so dark but what there is sunshine back. And Jenny I am sure there is sunshine in store for us. How I should like to have been with you in your walk on the ice. I have not seen any ice this winter to speak of. When it freezes here it is only a thin scab, and all disappears before the next night. My desk is near a window, from which I can view the camp occupied by our regiment a few days ago. How changed it looks. A few days past rose a village of tents, but now all is deserted and nothing marks the place but the cedar boughs we planted around our tents, the piles of straw we used for beds, and the railroad which we constructed to keep us from the mud. This brings to mind the words I have often heard, that change was stamped on everything. Here Jenny is where one can realize it. 'Tis getting dusk, the sun has found its bed in the distant west. The boys are getting their chores done up for the evening. A bright fire glows in the large stone fireplace. The doctor and his attendants are seated around, conversing in cheerful tones. The sick are very comfortable. Three have died since I last wrote. I think the most of the sickness is caused by overloading of the stomach. They have the measles, get better so they have a good appetite, and then eat everything they can get, which of course causes sickness. After they are taken the second time, they hardly ever recover. I must give [your brother] Darius credit for resisting temptation.[29] I have been placed in his situation myself. Tell your father I should like to have helped him in his contest with the hickory nuts. Yes Jenny if I could be with you and enjoy myself as in days past, it would be happy moments to me. Give my best wishes to your mother. I hope someday she will see why her Jenny loved a soldier. Goodnight dear Jenny, may the good angels guard you and bring me in due time to your arms. James.

Camp Dana,
Belmont, Kentucky,
March 14, 1862

Dear Jenny,

I received orders day before yesterday to leave Bardstown and report myself to headquarters at Belmont. I had received orders, and all I had to do was obey. My things were all packed, and I started for the depot. I found the train ready to start & all I had to do was step on board, the steam was let on, and I was whirling onward toward my destination. It was very early in the morning, the sun shining clearly. Onward we went and still onward, passing plantation after plantation. Stopping at stations every few miles on the road. We passed some very beautiful scenery in our route. Sometimes we were riding between rocks that rose far above us. (I would remark here that this road some of the way is cut through solid rock. It cost a great amount of labor.) At other times, crossing bridges over ravines which seemed, when looking out of the window of the car, as though we were flying. After stopping at several stationhouses we came to one, the conductor shouted Belmont Station, here was where I was to leave the cars. I seized my knapsack and accoutrements and started for Belmont Furnace, which was about 3/4 of a mile from the stationhouse. Well, all I have to say about my walk is that if you had met me with my knapsack on my back and all the accoutrements a U.S. soldier is obliged to carry, you would have taken me for a son of the Emerald Isle. I found Belmont Furnace, a place of about 50 houses, situated in a valley of about 200 acres, surrounded by mountains which raised their lofty heads far above us. These mountains are covered with a fine growth of timber. A few of their tops are covered with evergreens, which give them a very beautiful appearance. There is something of awe and grandeur in one of these huge piles [that] one feels which I cannot describe. Oh Jenny I wish you were here to look for yourself.'Tis a sight worth beholding. The principal buildings in Belmont is a large iron foundry, a large frame dwelling, and a large brick building toward which my steps were directed. Just as I was crossing the creek which comes down from the mountain, I met Lemuel with his team. He told me that he had been left behind to drive team and our company had gone to Elizabethtown, about 16 miles from here. I went into the building occupied by the officers and found him [Colonel May] in his room. He wished me good morning, asked me how the sick were getting along up at Bardstown. I told him that four of them had died since the regiment left."My God," said he, "how many of our boys are going to die? Would to God I had never seen a regiment." A great many blame Colonel May, but I believe if there ever was a kind-hearted man and one who tries to do right, Col. May is one. I have seen better managers than him, but he has a heart. He told me he wished me to act as clerk in the general hospital.[30] Our hospitals are in 24 buildings, which used to be tenant houses. There are patients here from almost every regiment which has crossed the Ohio. My duties here are nearly the same as at Bardstown. The two doctors, the steward, ward master, & myself occupy one building. We have everything handy and comfortable. From out of the window where I am writing, I can see the building occupied by the colonel, the bridge which crosses the creek, the iron foundry, and far in the background rises mountains. I can see the soldiers lounging around the streets. And one to gaze upon the scene would think it a quiet country town, that no war spirit hung over the place; but to look again, we see a sentinel pacing his beat on the road that leads to the stationhouse, also one in front of the headquarters. A daguerreian car also appears upon the scene.[31] There Jenny, I have just been making a furlough for a soldier belonging to the 50th Regiment Indiana Volunteers, it was nearly as long as this letter.[32] I have made 12 since yesterday morning. They are mostly boys who never ought to have enlisted. 'Tis amusing to see the poor fellows when receiving them. The time extended is thirty days. There are only two companies here, the rest are divided around in the other towns near here. One thing I can say, Jenny, is that we have the longest regiment in the service. It is scattered over nearly forty miles of ground. Give my best wishes to your father and mother, & believe me ever your James.

———————

The only extant letter from Jenny to James follows. Surviving nineteenth-century correspondence is often one-sided in this manner; women were more prone to preserving letters than men.[33] Naturally, this disparity in surviving correspondence was further widened in the case of soldiers, who necessarily minimized the bulk of their possessions. Additionally, as we saw in James's letter of November 6, 1861, Jenny on at least one occasion asked him to destroy a letter from her.

————————

[From Jenny to James:]

Fabius, Michigan,
March 23, 1862

The morning was clear and beautiful. But few dark clouds were visible, and having been requested to call at Mr. Walton's for the purpose of copying the family record, I at an early hour in the day started in that direction, father and mother accompanying me.[34] We had been there but a few moments when we were surprised to find it raining quite hard. But the storm was of short duration, for presently the sun again shone bright. Not wishing to remain long, I commenced my writing. It was soon finished. I told them that I thought we had better return home soon, as it looked likely to rain. And after a short and pleasant ride on the lake, [we] reached home. It was now two o'clock. I had resolved to write early today, and after supper. Clarinda is here yet.[35] But few of the pleasant spring days so earnestly looked for has come here yet. Yesterday was inspection day, it rained most of the time while we were going, but not very hard. Rebecca, Letitia, and Elizabeth were there. Emma & Henrietta [Vanderbilt?] went from Mr. White's.[36] I went with them, not knowing when the exercise commenced. We did not go until after noon and returned at six o'clock, having but five hours of recitation while most of the others answered questions from nine o'clock in the morning until six in p.m. I think the teachers of Township of Fabius have made great improvement in the last year. Last spring the inspector thought four hours sufficient time for inspection to last. I received your letter last night. I have wished that you might never have to leave the little village of Belmont unless to return home: but whatever changes there are in your regiment, the hope I cherish of soon seeing you again shall in part make me more patiently wait your coming. I should like to have been with you and watched the setting sun. I love to view it as it is sinking in the far off west. I love this hour. It is this hour more than any other that brings to my mind sweet recollections. But could I have been with you that night, the pleasure of your company dear James would have given me more and greater enjoyment than the sun in all its beauty and grandeur. There is but little sickness around here. I hope you will write as often as is convenient. Awaiting anxiously your return, I still remain your Jane.

Father & mother send their best respects to you.

Belmont, Kentucky,
March 23, 1862

Dear Jenny,

This morning when I arose, I made three resolves. First, that I would do all my work, second, that I would climb one of the mountains which rear their lofty summits far above the little village of Belmont. And third, write Jenny a letter. After breakfast I went to work. I had four requisitions to make for provisions, besides my books to post. By dinner time my work was all done, and the rest of the day was my own. Doctor Byran wished to accompany me, and immediately after dinner we started in our undertaking, which was no small task.[37] We took the road leading out of the town, and after a short walk, found ourselves at the foot of the object which we were going to surmount. This, Jenny, made me think of the hill of science we used to talk about being so hard to scale. While at the foot, I gazed up its steep sides and thought of the work it would take to reach the top. The sides were very steep, and 'twas difficult to ascend. But after hard walking we reached the top. Here we were repaid for the trouble we had taken. The top was covered with solid rock, and as I stood upon its solid summit and gazed upon the scene before me, my heart was filled with strange emotions. At its foot lay the little town of Belmont, whose small tenant houses looked like so many bird cages. To our right, at the foot, ran the road in its serpentine course among the hills. The valleys were green, and the brown sides of the mountains contrasted strangely with them. Far beyond the town, toward the Ohio, rose another range of hills, their tops, many of them, covered with evergreens which seemed to mingle with the clouds. After we had enjoyed the scene a long time, we made up our minds to descend. We were not as long in gaining the valley as we were in ascending. Just as we struck the road, we met [Assistant] Surgeon Whelan & [Lt.] Col. Stoughton on horseback. They galloped off up the road leading among the hills, and we returned to our room. We found a good warm fire in the office awaiting us. Dr. is reading a paper, & I am. I will leave you to quip. The boys are getting supper in the other room. They have great times in performing their household duties. One of their names is Joe, the other Jake. Joe is the greatest fellow I ever saw. Day before yesterday, he sent a couple of army crackers neatly done up with a lot of putty on them to a girl. He came to the desk, got me to direct it, [with me] thinking that he was sending his likeness to some fair one. After I had directed it he told me what it was. He wrote the following words and sent with it. "Enclosed I send you a specimen of the substance with which we are furnished." I have written this to show you what tricks are passed in the army. These fellows were present at the killing of John A. Washington last summer in Virginia. John A. Washington was the last of the Washington family, or the last relative of those who bore the name. They were in ambush in the mountains of Virginia; near them ran a road which wound around among the mountains. They saw approaching them six Rebel officers mounted on white horses. The Rebels perceived them, and one of them fired a revolver at them, at which they whirled their horses and started back in retreat. The boys fired, emptying two saddles. One of the officers was killed instantly, and Washington mortally wounded. They bore him to their camp, where he expired. 'Tis strange that one bearing so honored a name should prove a traitor to his country.[38] I have just finished my supper and have seated at the desk again, and as I raised this sheet and turned its page, methought how soon thy own hand would perform the same act. Thy sparkling eye scan the words that I am penning, and I so far from you. What a glorious thing is the silent language of the pen. It has caused my heart to leap for joy many a time since I have been in the crowded camp, and caused my duties to grow light. The shades of eve are settling on the Earth's broad breast, and as she weaves her dusky shadows round each object, methinks how oft I've done the same in my own dear home. Could I be with you this evening, it would be a joy to me. But fate have it otherwise. I ofttimes find myself pondering o'er the scenes of my life, and think how strange it has been, who would have thought one year ago that I would be here in my present situation, a U.S. soldier. But the ways of this world are many. Jenny I have seen many painful sights since I have shared the life of a soldier, and had many joyful times which will be pleasant to reflect on. But there never was a sweet unless the bitter was mixed with it. I have found many men in the army who make noble friends, trustworthy companions, and who are willing to aid one. But the army could never keep me if we were not contending for one of the noblest causes that mankind were ever engaged in. War is dreadful, and to me it would have no enticements if it were not as it were. But dear Jenny it will not be long before sweet peace will brood over the land and prosperity will again bless us. Joe the cook has just come in to get a discharge; he wants to go to the regiment. The doctor tells him he don't want to discharge him, he likes his style. He has begun to talk and I must stop, there is no writing where he is. Write soon. Remember me as ever your James.

Excuse poor writing as I have done this between light & dark.

Enclosed I send you a specimen of crystalized quartz.

Belmont, Kentucky,

March 30, 1862

The day has been passed by me very pleasantly, more so than any I have spent in the army. I have spent most of the day roaming over green fields and in the open air, for 'twas too lovely to stay indoors. In my rambles today dear Jenny, how often have I wished that you could enjoy the scenes with me. I visited an old plantation house in my rambles, which was deserted. As I roamed through the deserted rooms and gazed on its vacant walls, my thoughts were busy with themselves. The building was one of logs and is what with us would be called a double log house. There were six rooms below, a hall led through the middle of the house, and from this hall were doors leading into each of these rooms. The windows were boarded up, the walls were ceiled with boards which made it dark and dreary. As I gazed on the dark and gloomy walls, methought perhaps could they but speak would tell me strange stories. Perhaps the gay of life had trod these floors, and mirth and jollity had once been visitors here. But now how changed. This habitation once tenantable, now deserted. Nothing will set my thoughts to work so quick as a deserted dwelling. Thoughts, how strange; what a world of mystery is encompassed in this small word, "They know no distance, no time, no obstacles. Mind is their birthplace, burial, they have none." How often in thought have I visited you today, and yet so far from you. If 'twere not for thought, what would we be, dear Jenny? "Though they move mind by enchantment, yet the changes effected in the outer world are but the faintest echoes of those in the inner, are but the dim starlight that tells us of a universe. No sparkle of beauty nor voice of life decked or gladdened the new Earth until the Sun drove back its waters and developed its germs. Lo, mind is without form and void until sunlight from nature's characteristics, slimes in through the senses, developing bright blooms of first conceptions." But I was telling you about this house and nearly lost myself. From the house I visited the orchard. Here I found neglect and degeneration. But then I found relief gazing at the green and tender herbage at my feet. Nature's pictures are always more pleasing to me than arts. And if one wishes to see nature in her many curious forms, he should visit Kentucky. Dear Jenny I must thank you for your letters.

I received three from you last week, and they were fond sources of enjoyment to me. I should have answered them before this, but I have been very busy & I wrote your father & mother one, which made me more negligent. Dear Jenny how often I think of you. I still remain your James.

You wished to know if Lem got intoxicated. I saw him so once, but do not mention it to anyone. I think he will do better. Excuse mistakes.

Belmont, Kentucky,
April 6, 1862

Dear Jenny,

    I heard this afternoon that we were brigaded, and if so our stay here is short. Colonel May has resigned, at least I have been told so by authority. Who will be our next colonel I cannot say, but there is not much doubt that the lieutenant colonel will take his place.[39] I think Lieut. Colonel Stoughton a better military man than May, and the regiment will probably do better under his management than May's. We are going to have new guns furnished us this week, and that goes to show that our stay is short. The regiments that we are to be brigaded with are below Nashville. The day has been very beautiful. Could you but have been with me this eve to see the orb of day set in his golden bed, it would have been a sight well worth enjoying.

> Confessed from yonder slow extinguished clouds,
> All other softening, sober evening takes
> Her wonted station in the middle air;
> A thousand shadows at her beck. First this
> She sends on earth; then that of deeper dye
> Steals soft behind; and then a deeper still,
> In circle following circle gathers round,
> To close the face of things, a fresher gale
> Begins to wave the wood, and stir the stream,
> Sweeping with shadowy gust the fields of green.[40]

    Yes Jenny if 'twere not for friends whom I have left behind, and the love I still cherish for them, it would be happiness to me to enjoy these beauteous things. But nothing is there in this wide world that could call my thoughts from Jenny. What are beauteous things compared with those we love? Yes Jenny I would give all coming enjoyments, could I but be with you this evening and hear that loved voice once more. 'Tis 11 o'clock and the boys have retired. I sit by the desk, my seat a box, my light a candle. And Jenny I must confess, the candle is nearly burned out, and if I go upstairs to get more, [I] will wake the boys who would joke me tomorrow about sitting up all night to write a letter. Therefore I will close this by wishing you a kind goodnight. From him who thinks of you often. James.

Belmont, Kentucky,
April 14, 1862

Dearest Jenny,

I received your letter on Thursday, and to me it was truly a welcome guest. Your letters, Jenny, have this about them. "I never find anything in them but what gives me pleasure; for which therefore I would take nothing in exchange that the world could give me, save and except that for which I hope to exchange soon (and happy shall I be to do so), your own company." Yes Jenny how I would like to have been with you in the walk you mentioned, and would you not [have] considered it strange if I had not accepted the invitation?

The scene which has kept the national heart pulsating wildly for many days has at last broken over us, but with a dreadful loss of life on both sides.[41] What a scene to be enacted on the Sabbath day. I suppose you have seen a detailed account of the battle before this time, and as you read, almost hold your breath in suspense as you recoil from the stunning reports. "And oh, the agony and desolation that will come to thousands of homes in the noble North and in the erring but still beloved South" as the reports reach them of fallen friends in battle. This, Jenny, was one of the most dreadful things that has transpired in modern times. All the others have been mere play to this. The taking of Island No. 10 in the Mississippi River would have caused the greatest of exultation, had it not been for this last great victory.[42] The nation was looking to this great battle, and in the excitement, Island No. 10 was almost forgotten. I do not think the Rebels will make another stand unless it be at Yorktown, where they may probably risk a battle.[43] The bravery shown by our Northern troops truly deserves praise, and shows well their feelings in the noble cause in which they were engaged. And it must be confessed, the Confederates fought like men worthy of a better cause.

There is nothing new transpiring in the regiment; only a little excitement as to who will fill the vacancy caused by May's resignation. But there is no doubt but what our lieut. col. will be chosen colonel, and who will take his place remains for the future to decide. One thing, I must give Col. Stoughton the praise. He has exchanged our old muskets for the latest style of improved Springfield Rifle.[44] I have not seen Lem in a long time, but I hear he is well. Cuthbert Dixon has been very sick. I am told he is now getting better.

There is no one in the office but myself. Just now, "I am monarch of all I survey, my right there is none to dispute."[45] No one to come to the desk and try to find out what I am writing. We are going to break up the general hosp. at this place next Thursday, and then I will go back to the company.

Give my best wishes to your father & mother. Write soon and believe me ever your James.

Belmont, Kentucky,
April 25, 1862

Dear Jenny,

You probably have looked anxiously for a letter from me, and of course been disappointed. The reason of my not writing was this. I have been confined to my room nearly a week with a severe fever, but am now able to be around, though not to perform my duties in the office. Our regiment has orders to move to Nashville tomorrow. I do not know whether I will go with them or not. If our surgeon keeps charge of the general hosp. at this place, I will remain with him. The rest of the boys are well. I received two letters from you since I have written, and truly Jenny, they were a fond source of pleasure to me.

I cannot write any more at present, my hand trembles. Goodbye Jenny, may the good angels keep you. From your James.

In the aftermath of the bloodbath at Shiloh Church, the Union army pushed southward, and the 11th Michigan was pulled along in its wake. On April 28 James and the regiment departed for Nashville, proceeding by rail to Louisville and then by river steamer to their destination, which they reached on May 2.[46] For the first time, the Michiganders found themselves surrounded with a hostile population.

CHAPTER 3

The Celebrated John Morgan

May–December 1862

―∞―

I never saw a more exasperated set
than our men and officers.

Unionists looked forward to the military operations of 1862 with abundant optimism. In the East, Major General George Brinton McClellan engaged in a campaign widely expected by Northerners to conclude with the capture of the Confederate capital. In the West, the victory at Shiloh had paved the way for another Federal advance. New Orleans succumbed to a fantastic naval assault led by David Glasgow Farragut, leaving Vicksburg as the only major obstacle to Federal dominance of the Mississippi River. It seemed the Confederacy would soon be cleaved in two and its seat of government conquered.

At Nashville, the men of the 11th Michigan occupied true enemy territory. The Wolverines had left behind the comforts of civilian hospitality, but the soldiers were anxious to finally take an active role in the war. Spirits ran high. Confederate general Braxton Bragg would soon set his sights on Kentucky, and the 11th's position astride the Federal supply line placed the regiment right where the enemy cavalry would aim to strike. In the months to come, famed Rebel raider John Hunt Morgan would render the Michiganders harried and frustrated.

―――

Nashville, Tennessee,
May 3, 1862

Dear Jenny,

Here I am in Nashville, one of the most beautiful cities in the Union, and as I pass along its fine streets and gaze on its palaces surrounded with trees of the richest foliage, one would hardly think it one of the hottest beds of rebeldom. One can hardly think that a people surrounded with every comfort of life, and living under the best government in the world, could become so dumb to their own interests as to turn traitor to their country. The tombs of the illustrious patriots who sleep near this place ought to remind them of loyalty. The shadow of the tomb of the illustrious patriot [Andrew] Jackson ought to cause black-hearted secessionism to shrink back to the place from whence it started and never show itself again.[1] Our trip down the Ohio was a pleasant one, and will never be forgotten by me. From Louisville to the mouth of the Cumberland, the country was entirely submerged, a distance of over three hundred miles, and looked more like a vast lake than a river. Whole farms were entirely under water, and the people were obliged to use the upper stories of their buildings. I tell you it looked dreary enough, most every building had a skiff tied to the door latch. But the flood did not stop their cheering us. The night of the second day, we reached the Cumberland. This river is about as large as the St. Jo, only a great deal deeper.[2] I remained on deck as long as I could see anything, and then retired to the cabin. I was rather unwell, and Doct. Whelan furnished me with a stateroom. In the morning I was on deck as soon as 'twas light enough to see. We were still going at the rate of 8 miles an hour. The scenery on this river surpasses anything I ever saw. The banks, which rise hundreds of feet from the water's edge, and covered with trees of the richest foliage, makes a scene that is lovely to gaze on. At 8 o'clock all was excitement, as we were soon to pass Fort Donelson. The fort soon came in view. Here was where one of the bloodiest battles of the war was transacted. Close to the water's edge was a breastwork where they had several batteries of the heaviest caliber. These were the batteries which done so much damage to our gunboats. Farther up the bank, where the battle raged with the greatest fury, was another breastwork. I noticed some trees which looked as though ten thousand lightnings had played among their branches. 'Twas through these trees our gallant boys poured their shot and shell, sending death and destruction into the enemy's camp. A little farther up, we came to the iron foundry of John Bell, which our troops destroyed.[3] 'Twas burned by our troops after the taking of Donelson. 'Twas used by the Rebels in casting cannon and rolling iron plating for gunboats. About noon we reached Clarksville, where the Rebels retreated from Donelson but left on short notice at the approach of our troops. They tried to destroy the railroad bridge that crosses the river at this place, but were unsuccessful. We passed on up the river, and just as the shades of evening threw her dusky mantle over the face of the earth, we came to Fort Zollicoffer.[4] It is situated in a bend of the river, having a good command either way. Three large cannon remained on the walls. I saw the end of one sticking out of the water. The Rebels upon leaving sank a number of their cannon in the river. We arrived in Nashville about 10 o'clock in the evening, and remained on the boat all night. The people here are nearly all secessionists, and look upon us with a coldness. The nicest country I ever saw is here. Our camp is in a grove of trees. The land is owned by Senator Ewing, who once occupied a seat in the United States Senate but is now in the Rebel congress.[5] His wife remains here yet—a Negro told me this morning that he guessed his missus would go crazy. His residence is near here, and is a perfect palace situated in a fine grove of trees. The yard is filled with flowers of every description. The weather is very warm, but pleasant. Lem is in the tent and has just borrowed some paper of me to write a letter. Cuthbert Dixon is also here but is unwell, he has been very sick. There is some prospect of our moving farther down in Dixie. I do not think we will stay here many days. The Ohio 51st Regiment went out the other day about 25 miles from here and captured 300 Rebels, among them was a lieut. colonel & 3 or 4 captains.[6] They are now confined in this city. There is some talk of our going to Corinth, but no certainty. My space is getting short and I must close. Wishing you the best of success in the trials of school teaching, I still remain your James.

Give my respects to your father & mother.

Nashville, Tennessee,
May 6, 1862

I was starting to take a stroll with a friend when Dixon brought me your letter. Of course the first thing I done was to read it, and now accompany me in my stroll. It is about 4 o'clock in the day. The Sun shines pleasantly. We pass o'er a green lawn shaded by noble trees. We are going to view the garden of Senator Ewing, who owns the farm on which we are encamped. A high stone wall surrounds it. We enter a gate and walk along the clean and well kept paths; with roses, snowballs, and shrubs of every description on every side. There is probably an acre of ground in this garden of flowers and ornamental shrubs. Here I saw the famous fig tree flourishing. The magnolia moving its broad leaves in the breeze. Well, we have went almost around the place and now we will take our leave. There, I have picked a rose and a sprig of the juniper tree for you. We pass out, but our attention is attracted by a burial place near the garden. There are several graves or tombs built of brick on the tops, covered with marble or iron. Now I wish to copy one of them. I have a pencil, but no paper. Yes Jenny I will take your envelope.

Sacred to the memory of Col. Joel Lewis, who was born August 28th 1760. He was an officer of the United States during the Revolutionary War, and contributed his full share of patriotism, prudence, & bravery towards the independence of his country. He reared a numerous, respectable family and set them a striking example of the life of a virtuous man, and useful citizen. He was a firm and practical believer in the doctrines of Revealed Religion, and with unwavering faith and Christian hope, he died on the 22nd day of Nov. 1816.[7]

The regiment has left Nashville, and left us behind to bring the tents and baggage. Keep up a good heart Jenny. A light heart goes merrily through the world. Goodbye. From your James.

————————

James rendezvoused with the regiment at Columbia, where the Michiganders had deployed in response to false reports that Morgan was operating in the vicinity. Intense guerrilla activity kept Stoughton's bluecoats on their toes around the clock. On June 2 Tennessee's military governor Andrew Johnson and former governor Neill S. Brown arrived by train to address the populace. Brown had opposed secession before the war but served the Rebel Tennessee governor, Isham Harris, in an administrative role after the conflict broke out. He was arrested on May 14 for treason but promptly took the oath of allegiance and joined Johnson's efforts in speaking out against secession.[8]

————————

Columbia, Tennessee,
May 16, 1862

Dear Jenny,

Again have you been disappointed on account of my not writing, but the same power which delayed before has been at work this time. When I reached this place, I sat down to write you & sister a letter. I finished hers, and was commencing yours when I was called away, and kept busy the rest of the day. The next day I was taken with fever again and have had a pretty severe time of it, but feel more like myself than I have since I was first taken.[9] We are camped on the side of a tall hill which overlooks the town. Columbia is a regular secesh place. Our regiment has done a great deal of scouting since we came here, and brought some prisoners into camp, but have not had a chance to fire a shot yet. The regiment leaves for Huntsville, Alabama.[10] The tents and a guard are to be left a few days, when the teams will return and get them. I received your letter dear Jenny, it pleased me much, and after all it pained me to think your thoughts were so gloomy. We all have our trials to bear, and let us bear them with a will. This war I think is near a close. Excuse this short letter, I will do better next time. As ever your James.

Give my best wishes to your father & mother.

Columbia, Tennessee,
June 2, 1862

Dear Jenny,

The shades of eve have fallen over the face of the Earth. The lights are lighted in every tent, which makes a very beautiful sight. Today has been one long to be remembered by me. There was to be a great Union meeting, and at an early hour the people began to flock in. Men who had not been in town since our troops gained possession of the town. I could tell every one of them by their shy looks. At 11 o'clock the train came in from Nashville with Governor Andrew Johnson & Ex-Governor Brown, besides many other distinguished individuals. The 69th Regiment Ohio were on board the train.[11]

Our regiment were drawn up near the depot. As the governor stepped from the cars, the band commenced playing. They played several pieces. The troops were then formed into line and escorted him up town to the courthouse. A vast crowd of citizens followed them. The governor was escorted to the Nelson Hotel, where he made a few remarks to the crowd and then retired from the scene. About 1 o'clock the speaking was commenced by Ex-Governor Brown. He told the people that they were wild to think of holding out any longer. The first place, they had not the means to carry it on. In the second place, six millions of people contending against twenty-five millions was all nonsense. He said that he had two sons, one brother, besides many other relations, in the Confederate Army, but he could not help it. He never told them to go. He felt more like weeping than doing anything else. His brother was now a prisoner in a Northern fort. His two sons were in the disorganized army of Corinth, and where that was, the Lord only knew. Tennessee was his native state, and he loved her as his life.[12]

Governor Johnson next came on the platform, he was greeted by enthusiastic cheers. He came forward and said that he was pleased to see so large an audience present. He wished to talk with them on this all important question. For [a] full 4 hours he talked with them about their troubles, until nearly every one shouted for joy.[13] In the meantime, we had received news that the enemy would attack the town before night. Four pieces of artillery were placed at a fork of the road where the Pulaski pike or road came in. Our regiment was placed in ambush and waited nearly all the afternoon, but no enemy made their appearance.

About 5 o'clock the meeting broke up. The governor took the train for Nashville. Just as they were ready to start, the band struck up Yankee Doodle. He left amid vociferous cheers. We returned to camp. Soon after we got into camp, part of the regiment were ordered out on the Pulaski road to support Capt. Stanard's [Standart's?] Battery.[14] There is about 300 men in camp tonight. The colonel has just been here, and asked me if this tent was not a brigadier general's. I asked him if it did not look like one. I have received three letters from you since I wrote you. Day before yesterday I received 2, and today, one. I thank you for them. They have given me great pleasure. In one of your letters, you asked me if this fine country and pleasant scenes could make me forget my friends at home. Never! Dearest Jenny never! You may think me cruel and unjust Jenny if I should accept a position in the Regular Army, in the Quartermaster Department. I can get a good position, and they cannot hold me any longer than my term of enlistment. You may think I have forgotten you or do not care for you. Ah Jenny, absence makes the heart grow fonder. I think that it is [in] my interest to accept the position. Business after the war will be very dull, and a clerk's life in the Quartermaster Department of the Regular Army is one of the best positions in the army. I have just heard 2 guns of the pickets, but it must be a false alarm. We hear this most every night. We have got so used to it, that is hearing this firing, that it does not disturb us. The commissary is spreading our couch for the night, and I must close.[15] With the kindest regards for your father & mother, I remain your James.

Lem has gone to Pulaski with his team, with a load of provision.

[Columbia, Tennessee, early June 1862—initial page(s) missing—ed.]

Dear Jenny, in this sequestered scene,
While all around in slumber lie,
The joyous days which ours have been,
Come rolling fresh on Fancy's eye
Thus if amidst the gathering storm,
While clouds the darkened noon deform,
I hail the sky's celestial bow
Yon heaven assumes a varied glow,
Which spreads the sign of future peace
And bids the war of tempest cease[16]

I could not resist the present opportunity of writing you, though I have not heard from you in a long time. I do not think I get all the letters that are written me. The mails are very irregular, there are very few letters received in the regiment. We are still in Columbia, but have orders to be ready to march on a moment's notice.[17] The boys are all in good spirits. The weather is fine, though pretty warm. Wheat is all headed out and will soon be ripe. We have had strawberries for nearly three weeks. The season, I think, is all of 6 weeks ahead of Michigan. News to write is very scarce. I am in the Quartermaster Department, my business is principally writing.[18] Excuse this short letter. Give my best respects to your father & mother. Remember me as ever your James.

––––––––––

The regiment deployed to Murfreesboro by train on June 9. James and Lemuel brought the regimental supplies separately and did not arrive in time to rendezvous with their unit, which departed June 11 on an expedition under Brigadier General Ebenezer Dumont to intercept Rebel cavalry reported in the countryside. The 11th marched well over two hundred miles but still saw no action, and the regiment returned to the drudgery of garrison duty at Nashville on June 21.[19]

––––––––––

Murfreesboro, Tennessee,
June 15, 1862

Dear Jenny,

'Tis evening, "silence now is brooding o'er the still and pulseless world."[20] The noisy scenes of camp have ceased, and quiet reigns for once. 'Tis past the hour of 9 o'clock, the bugle has blown for lights out. The soldier has wrapped himself in his blanket, chosen his bed on Mother Earth, and is now reposing as quietly as though he were surrounded by the scenes of home, or beneath his own peaceful roof. At the close of the day, just before the Sun had sunk to rest, a companion of a fellow clerk of mine procured horses for a ride. We took the road leading into Murfreesboro. A ride of one quarter of a mile brought us to the village. The village itself is larger than Three Rivers, and has one day been a place of considerable importance, but its doors are closed and all business is suspended. Bevies of ladies and gentlemen thronged the sidewalks, a great many of them dressed in mourning. I thought as I gazed on these signs of grief how many once happy homes were mourning the loss of some loved one who had fallen a victim in this unholy war which is now raging in and around them. Ah Jenny, war is dreadful. We passed on through the village and soon found ourselves in the country. The fields of young cotton on either side of the highway attracted our attention. 'Tis planted in rows or drills about 3 feet apart, and is now about 7 inches in height. It now resembles buckwheat. We went down this road about 3 miles, then turned our horses' heads for camp. On our way back, we met several young ladies. (Here let me remark that some of these ladies; no, I will not call them ladies, have a way of showing their contempt for Northern soldiers by making faces at us.) Well, one of these as we passed greeted us in this style. I stopped my horse. Say I, "Madam, 'tis a blessing that you Southern women live in a climate where it never freezes, if so there would be many wry or crooked faces." She replied by a haughty shake of the head.[21] We turned our horses and galloped to camp. We reached this place after a march of 2 days from Columbia. I came through with a train of provisions. The regiment, by railroad. We failed to make connection with the regiment and [were] ordered to remain here while the regiment have gone on. Lem is here in camp. Several of the boys who failed in the march report that our boys have between 60 and 70 prisoners which they have taken on the road. The town clock strikes eleven, and I must close. May God bless you dear Jenny is the earnest prayer of your James.

Nashville, Tennessee,
July 3, 1862

Dearest Jenny,

I should have written you ere this but I have been very busy for the past week, so much so that I could not find time to write a letter. The quartermaster is settling his accounts, and I have hardly left the desk 10 minutes at a time in a week, though I did find time to visit the grave of the illustrious patriot and statesman Andrew Jackson, whose ashes now repose in the beautiful state of Tennessee. That visit I shall never forget. Last Sabbath morn was a pleasant one, and Lem, Dick, & myself hired a hack and set out on our journey. Along the road on each side were fields of corn higher than a tall man's head, waving in the summer breeze, also fields of cotton 6 or eight inches high. Then a piece of beautiful wood, added to the fine residences back from the road, surrounded by groves of trees of every description, made a scene too romantic for description. 'Twas nearly noon when the driver reined his horses into a crossroad, and after a short drive we found ourselves in front of the Hermitage.[22] Two carriage roads led to the mansion through beautiful groves of cedar trees. We alighted in front of the structure which was once the abode of him whom a nation reverenced and looked on with admiration. What a strangeness of thought crowded on my mind as I gazed on each scene. The boys wished to eat our lunch before doing anything else. A Negro had conducted us to the spring house a little back of the main building, and there, under the shadow of a protecting oak which spread its leafy boughs over us, we partook of our dinner. After dinner we took the path to the tomb, which was in the back part of the garden. I was struck with the plainness of the tomb, no show of grandeur or richness. 'Tis of an oval shape and built of marble. The roof is supported by four marble posts or pillars. The masonry was poorly executed and is now falling off. The simple inscription telling when born and died were all; 'twas enough, he needs no more. On his wife's grave was one very beautiful, which I copied. This is just as I copied it.

Here lie the remains of Mrs. Rachel Jackson wife of President Jackson who died the 22nd of December 1828 aged 64 years. Her face was fair, her person pleasing; her temper amiable and her heart kind. She delighted in relieving the wants of her fellow creatures and cultivated that divine pleasure by the most liberal and unpretending methods. To the poor she was a benefactor, to the rich an example, to the wretched a comfort, to the prosperous an ornament. Her pity went hand in hand with her benevolence, and she thanked her Creator for being permitted to do good. A being so gentle and yet so virtuous, slander might wound, but could not dishonor. Even death when he tore her from the arms of her husband but transport her to the bosom of her God.[23]

Jenny is this not beautiful, not a word could be added.

From the tomb we were invited into the house. The house is built of brick and is of the first class. Two piazzas run the length of the building, one above & one below. From the piazza we entered the hall. This contained several rich paintings, and in the corner sat the old armchair presented to Jackson by Washington.[24] I of course had to seat myself in it. From the hall we went into the sitting room, which was elegantly furnished. There were two paintings of Jackson, one of his wife, and one of the present proprietor, Andrew J. Donelson.[25] There was a wooden pitcher sat on the table with an inscription which was quite a curiosity.[26] After seeing all that was to be seen, we entered our carriage and drove slowly away from the grave of the old hero.

I received your letter day before yesterday, 'twas truly a welcome guest. I was glad to hear that you were enjoying your school so well. A light heart, Jenny, goes merrily through the world. I spoke of leaving the regiment in one of my letters. The situation which was offered did not need another clerk as was expected, therefore I will remain where I am. My health is as good as it has been since I enlisted. I cannot write any more. Goodbye. May the angels guard you. Yours forever, James.

My best wishes to your father & mother.

––––––––––

Buell's Army of the Ohio in the meantime was struggling through a march east from Corinth, Mississippi, toward Chattanooga, incessantly harassed by Confederate cavalry and guerrillas, who repeatedly severed Buell's line of supply. On July 9 John Hunt Morgan surprised and routed a detachment of the 9th Pennsylvania Cavalry garrisoning Tompkinsville, Kentucky. In response, the 11th Michigan proceeded to Cave City, Kentucky, where they learned that Morgan was presently at Glasgow. Stoughton trudged his men in pursuit through the mire of an overnight rainstorm, but their quarry eluded them. The regiment reported to Louisville on the morning of the fifteenth and then removed by rail to Frankfort, where a force was assembling under Colonel Cicero Maxwell to coordinate with Brigadier General Green Clay Smith, located at Lexington, in the pursuit of Morgan. The Wolverines joined a mixed command of infantry, cavalry, and artillery in an overnight march to Georgetown. Morgan had rested his troops there for two days, but true to character he slipped away shortly before his pursuers arrived. The Federals soon learned that Morgan had taken the town of Paris, and they caught up with the Rebel horsemen this time but failed to coordinate their forces to prevent the Confederates from retreating. Once again, Morgan made a clean escape, and the 11th returned empty handed to Nashville.[27]

––––––––––

Nashville, Tennessee,
July 11, 1862

I received your letter this morning, and it found a welcome reception. I am still in Nashville. The regiment have gone to Glasgow, Ky., to capture the celebrated John Morgan; and I remain behind with the quartermaster stores.[28] There is about fifty men left in camp. The regiment went from here night before last about 12 o'clock, by rail to Bowling Green, [and] thence to Glasgow.

We expect them back day after tomorrow, when we will probably go to Huntsville and join Buell's forces in Alabama. Everything is quiet here. After we left Murfreesboro, there were three of the Pennsylvania cavalry killed while scouting near that place. The 4th passed very quietly in camp. There was a celebration in the city, but I did not go down until evening. There was a Union speech delivered by Ex-Governor Neill S. Brown of this state, which was well received.

I am happy to hear that you enjoy your school so well. The teacher's avocation is a glorious one indeed, and it pleases me Jenny to think you love it. Those who labor faithfully Jenny will yet meet their reward. How I should love to be with you and sail over the waters of the blue lake where oft we've been before. I oft think of the happy times that have passed and are now remembered among the scenes of bygone days, but there is a bright future Jenny. When wars alarms are o'er, peace spreads her gentle pinions o'er this once happy land, then it will be sweet to return to those we love, but not until then dear Jenny. Not until our country is saved from the villainous hand of the scheming traitor. It may be a few short months, or long years. But let either come, I will not repair. Yes Jenny 'tis sweet to ponder and gaze in the future, to think of the happy days yet to be passed. 'Tis hope dear Jenny for the future. Memory the past. You remember Jenny reading the story of Hope and Memory traveling the land. They were jealous of each other. Each thought they caused more happiness than the other. But they finally concluded to travel hand in hand.[29] I cannot tell Jenny which brings me the most happiness, Hope or Memory. Write soon Jenny. You may think me abrupt in closing, but I have not time to write longer. Give my best respects to your father & mother. Yours forever, James.

Nashville, Tennessee,
July 16, 1862

Dear Jenny,

I received your letter yesterday and was surprised to hear that I did not write as often to you as I did to sister. Now Jenny I thought I wrote as often to you as I did [to] anyone. At least, I have written more than I have received, and I do not think I am to blame. For a few weeks past I have been very busy, that is one reason I have not written oftener. I will try to do better in the future. [Even] if I do not write as often as you think I ought, my thoughts towards you remain the same. This morning all is quiet, though considerable excitement prevailed among the citizens that an attack was meditated on the city. Reinforcements kept arriving yesterday, and by last evening there was a force sufficient to withstand any attack. Cannon were placed on every road leading into the city and several pieces of artillery were planted on capitol hill near the statehouse, but no Rebels came. This morning we received news that they were retreating to the mountains. I suppose you have seen the reports in the papers of the capture of

Murfreesboro by a body of the enemy's cavalry. The Michigan 9th were nearly all taken prisoners, and also the 3d Minnesota Regiment.[30]

This scene was enacted within 30 miles of us, and we had not sufficient force to lend them any assistance without endangering the safety of Nashville. They were surprised in the morning by about 3,000 cavalry led by the renowned Champ Ferguson.[31] Our regiment has gone to Louisville, Ky. How long it will remain there, I know not.

The boys are all well, excepting Wm. W. Hoisington, who is in the hospital sick, but not dangerous. We have moved our camp into the city. We were encamped out of the city about a mile, but were moved into the city when it was thought in danger. There is no danger whatever, as there is not sufficient force to make a successful attack. From your James.

Nashville, Tennessee,
July 30, 1862

'Tis a rainy night, darkness broods o'er the camp, the rain comes pattering on the tent, the bugle has sounded for roll call, and I can hear the boys answering to their names. These scenes are transpiring while I sit on a box by my desk, penning this note to you. The commissary & quartermaster sergeant are in the tent conversing on the prospects of the war, the principal topic of the day. They seem to think that if the quota of men called for by the president is raised, at once, and the war prosecuted with vigor, this war will come to a speedy termination. But on the contrary, [if] there should be a lack of energy on the part of the government, and foreign powers acknowledge their independence, it will take much longer.[32] They advance some good arguments. Both are intelligent men, and good reasoners. It is pleasing to hear them talk. When I last wrote you, we were at Bowling Green, Ky.[33] Our regiment joined us at [that] place, when we were ordered back to Nashville, where we arrived yesterday morning and went into camp. The part of the regiment who were in pursuit of John Morgan returned without capturing the enemy. At Paris, Ky., they came up with the enemy but were ordered to hold their peace and not fire on the enemy, and so allowed this black hearted traitor to escape. I never saw a more exasperated set than our men and officers. General Smith was the officer in command. A more dastardly piece of cowardice or traitorism, I do not know which to call it, has not been transacted in a long time.[34] I received a letter from sister complaining bitterly that I did not write every week, when I had not received one from her in over a month. Now Jenny I do not think she ought to have four for one. Maybe four of my letters are worth only one of hers, if so, we are just even. The boys

John Hunt Morgan's men in Paris, Tennessee. (*Frank Leslie's Illustrated History of the Civil War*)

have retired for the night, that is, chosen their beds on the ground to sleep until morn, when the rattling of the drum and the shrill fife will wake them from their slumbers. Holbrook says I must not write so long letters, but make them short and sweet. I might write short ones but the sweet part, that is what troubles me. The boys are well, except Wm. Hoisington, who is very sick of typhoid fever. Lem is well and full of mischief as ever. I think I will take Holbrook's advice, so goodnight dear Jenny. May we soon meet. James.

———

Morgan was ordered to break up the railroad between Louisville and Nashville to sever Buell's line of supply in preparation for Braxton Bragg's invasion of Kentucky. Morgan arrived in Gallatin at daybreak on August 12, captured most of the Union 28th Kentucky, and collapsed an 800-foot railroad tunnel outside of town. Morgan's main body departed before the 11th Michigan arrived the next morning. However, a Confederate detachment remained behind to torch an amphitheater, and the 69th Ohio, with the 11th Michigan in support, stumbled upon these Rebels at a cornfield near town. The Federals charged, but much to the Michiganders' chagrin, the Confederate horsemen fled. The Rebels later returned to ambush the bluecoats as they boarded a train.[35]

———

Nashville, Tennessee,
August 15, 1862

I was sitting and pondering just now what I would do, when the thought occurred to me to write Jenny a letter. No sooner thought, than I seized pen and paper and have commenced. Now what shall I write? I have nothing to tell you new, only, we have had a small battle with John Morgan at Gallatin, Tennessee, killing between 25 or 30 of his men, the loss on our side was one man killed and 2 wounded. But you say you do not like war news, pardon me Jenny if I tell you how this happened. News reached us the night of the 12th of August that Morgan had taken the village of Gallatin, torn up the railroad track, burned the bridges, and destroyed things in general. We left camp the night of the 12th at precisely 11 o'clock, went to the Chattanooga depot where we found a train awaiting us, loaded with the 69th Regiment Ohio Volunteers [and] a part of the 4th Indiana Battery, who were to accompany us. The night was splendid, not a cloud, the moon shining brightly. The train did not run very fast on account of there being danger, obstructions on the track. Just at daylight we reached the bridge burned by the Rebels. This was two miles this side of Gallatin. It took some time to unload the horses and battery from the cars, and by the time we were ready to start, it was nearly 7 o'clock. We took the turnpike leading to Gallatin and had proceeded nearly a mile along this road when our advance came unexpectedly on a detachment of the enemy. Our boys fired, killing three of them, 2 of which were captains. One of the Rebels had his horse shot under him, whom we captured. The rest of the Rebels put spurs to their horses and were soon out of sight. We entered the village and found it nearly deserted. The stores and houses were all closed. Orders were given to search every house. Stores were broken open, the boys helped themselves to whatever they wished.[36] We found several guns in the houses, all of which we took. We also recaptured a wagon load of guns taken by the Rebels from Colonel [William P.] Boone's regiment [the 28th Kentucky], who were made prisoners the day before. We remained in town nearly all day, when at 4 p.m. in the afternoon we returned to the cars. We were nearly all loaded when the guerrillas opened fire on us from a cornfield in which they were secreted, killing one man, 69th Ohio, and wounding 2 more. Our boys were soon out of the cars, pouring in the musket, grape, and shell. One shell burst in a squad of Rebels, killing four. They fired several volleys at us, but only the first took effect. We poured several well directed volleys into them, killing between 15 or twenty, when the Rebels scattered in every direction.[37] We then took the train car and reached the city

Timothy Robbins Stanley.
(Gracie, *The Truth about Chickamauga*)

about 11 o'clock at night. Wm. W. Hoisington died this afternoon. His disease was typhoid fever. His father was just in our tent, he feels very bad.[38] I do not know whether he will be sent home or not. How the folks at home must feel is better imagined than described, when they hear of his death. Ah dear Jenny, how many homes are made desolate by this unholy war. And many more will be made so before it ends. I wish you would write and let me know how many have enlisted, since the new call for troops, of my acquaintances. 'Tis getting dark and I must close. Goodbye. Write soon, yours forever, James.

––––––––

With Nashville's line of supply and communication severed, Braxton Bragg's ambitious invasion of Kentucky was fully in motion by the end of August. The 11th Michigan marched to Murfreesboro on September 1 but three days later was ordered back to Nashville.[39] After Buell passed through in pursuit of Bragg, the city garrison, under Brigadier General James Scott Negley, was isolated and surrounded with a hostile population. There were constant rumors of attack, and desperate foraging became a way of life. The 11th Michigan and 19th Illinois were combined with the 18th and 69th Ohio to form a new unit, Colonel Timothy Robbins Stanley's brigade of Negley's division.[40]

James Scott Negley.
(Gracie, *The Truth about Chickamauga*)

Bragg would retreat from Kentucky following his strategic defeat at Perryville on October 8, but Negley's garrison remained in isolation until lead units of the Federal army returned to the city on November 7. The corresponding interruption in mail service explains the lengthy gaps between James's letters.

Prior to the army's return, Buell's shortcomings resulted in his replacement with Major General William Starke Rosecrans. Rosecrans regrouped the army in and around the city, and camp life gradually returned to normal.

———————

Nashville, Tennessee,
August 30, 1862

Dear Jenny,

Today is a busy one in camp. We are preparing for a long march, in what direction I am unable to say. There are three regiments of infantry, 2 of cavalry, & 3 batteries to accompany us. There is nothing new in camp or in the city, except that troops are moving in every direction. We expect to leave the city this afternoon. There is a rumor of our going to Cumberland Gap, but there is no certainty. The Federal forces at that place are in a critical condition, and that is why I think we are going there. I cannot tell Jenny when this will reach you, as there is no mail between here and Louisville. Several bridges are burned on the railroad. I should have written before, but I knew they could not reach you. You must excuse this short letter as I have much work to perform. The boys are all well. Give my best respects to your father & mother. Farewell until I write again. James.

Nashville, Tennessee,
September 10, 1862

Since I last wrote, we have been to Murfreesboro and back again to this city. We received orders to move to that place the last day of August. When we reached the place, we found the whole of General Buell's force falling back on Nashville. We went into camp on the same ground the Michigan 9th occupied. After remaining here a day or two, we returned to Nashville, where we are at present encamped. There is now about 30,000 of Buell's army in and around the city. It is intended this force [Negley's division] shall hold the city while the rest of the army has gone to Kentucky. The reason of this falling back is this. General Buell has been lying before Chattanooga in a state of inactivity, letting a large force invade Kentucky and cut off his supplies. I believe Jenny, if the truth was only known, General Buell is a Rebel at heart and has done all in his power to aid the Rebel cause. His very acts show him to be one. I will say no more of him at present, but it will someday be shown to the world that Maj. General Buell was not a man to be trusted.[41]

I received four letters from you in one day last week, and now dear Jenny, accept my kindest thanks; I had looked long for them. You hoped I might be with you this winter. I hardly think I will, but there may be such a thing. The prospects of the war look gloomy at present, but I believe in a few short months this dark cloud will rise, and sunshine follow. The people of the North are arising in their might and waking up to their duty, and will sweep rebeldom before them like chaff before the wind. It is as you say Jenny, there has been too many officers who care more for pay than doing service to their country. But that day has passed.

I commenced this letter in the forenoon, but had to attend to issuing rations and postponed it. But now I am seated and writing again. We have our quarters in a private dwelling vacated by a secessionist. 'Tis a very pleasant place. Most of the officers occupy the dwelling; the tents of the men are in the dooryard and garden belonging to the premises. The dwelling stands in a low valley surrounded by 3 rises of ground; on each of these hills are forts. These forts are made after Hickory Jackson's plan of forts. They are made of cotton.[42] There must be at least one thousand Negroes employed on these forts. 'Tis a novel sight to see so many of these black men toting cotton.

We sent a party of foragers out this afternoon on one of the roads leading out of the city. Lem was among them. When they were out about 4 miles, 3 of the wagons were attacked by a party of guerrillas, and I think would have captured them if our cavalry had not come to their rescue.[43] Our boys captured one of them. Lem has just left the room. He is tough as ever. He wants to know if I am writing home. I told him yes, he wished me to give the folks his best respects. All the boys are well. I had a letter from brother John stating that he had enlisted.[44] Well I am not sorry. This rebellion must be put down if every Northern man is called into the field. I tell you Jenny we have been coping against fearful odds. In none of the battles which have been fought have we had a force equal theirs. Why did they not call out a million men in the first place instead of 75,000?[45] But hoping the government has awakened to the real dangers which now beset it, and traitors receive the penalty due them, I remain as ever your James. Give my respects to your father & mother.

Nashville, Tennessee,
October 29, 1862

After a silence of nearly 2 months, I again address you in the silent language of the pen. This is the second letter I have written since August. I wrote one & sent it with Dr. A. F. Whelan to my folks.[46] Whether it ever reached them yet, I am in ignorance. We have been, dear Jenny, completely isolated from all the world without; no mails, not a ray of intelligence reaching us, only that brought by couriers on which we could not depend. We are forced to depend on ourselves for everything. Our provisions have nearly all been foraged since the latter part of August. We have seen some pretty hard times, but most of the time have had enough to eat.

There is no news or nothing new to us. 'Tis one thing over & over. Every day. Here we have been since Buell passed through here, in the same camp, nothing going on except building fortifications & foraging. Our boys have skirmishes every day or two with small parties of guerrillas. But no heavy battles. Our regiment has lost none killed or wounded since we came into service. There were two of our boys taken prisoners and the probabilities are that they were hung.[47]

News has reached us today of the removal of Maj. General Buell and the gallant Rosecrans appointed in his stead.[48] This is hailed with joy by all the officers & soldiers in this division. The men & soldiers have had no confidence in General Buell as a commander for a long time.

We are having pretty good times here. If you could see our tent this evening, you would see Billy Davis & myself writing letters.[49] Billy cooks for us. There are two in the mess beside myself. The commissary & quartermaster sergeant are talking & laughing with a lot of the boys who have come into the tent to spend the evening.

I intend to send this by one Garrett, a QM of the U.S. Army, who starts for Louisville tomorrow morning with a flag of truce.[50] Oh how I wish the mails were regular. This going without letters when one has been accustomed to receiving them is very inconvenient, at least that is what we call it.

Give my best respects to your father & mother. Goodnight, may God bless you Jenny. Ever your James.

Nashville, Tennessee,
November 29, 1862

Dear Jenny,

Your very welcome letter came to hand yesterday evening, & I now hasten to reply. You cannot imagine the joy it gave me to receive a letter from you after so long a silence. I received a letter from sister dated Sept. 8th, stating that you were dangerously ill. Since that time, I had not heard a word until I received yours yesterday evening. I think there must have been some mistake, as you did not even mention it. Yes Jenny, since the reception of her letter I have lived in suspense for two long months.

There is nothing new or exciting in camp. We are here in the city and shall probably remain some time, at least until we are paid off.[51] Gen. Rosecrans has his headquarters here. General Sill's & Sheridan's divisions have moved southward & yesterday had a skirmish with the Rebels near La Vergne, a small town 15 miles south of this city.[52] He came upon the Rebel pickets this side of La Vergne. The Rebels fired a few shots and then fled until they fell in with other outposts, where they made another stand. After another sharp action the Rebels fled into La Vergne. It is reported that a very heavy force of Rebels are concentrating at Murfreesboro. There must be some truth in the report. Our generals are making great preparations for a large amt. of work.

The weather is rather cold, though not cold enough to freeze. Norman Hoisington stayed with me last night. He belongs to the 13th Regt. which is encamped on the Murfreesboro railroad. It was a great pleasure to meet him and talk over old times.[53] He started to go to his regiment about an hour ago. They were with Gen. Buell in his chase through Kentucky after Bragg. You spoke of the removal of Gen. McClellan & Buell. Buell's removal is hailed with joy by all. Some are dissatisfied with the removal of McClellan.[54] Think we ought to have tried him longer. But Jenny we have tried such generals too long.

This standstill policy I do not believe in, never did. I hope our new leaders will not follow in the tracks of their predecessors.

I received a letter a few days since from brother John in Kentucky. He is in the 96th Regt. Illinois Vols. He thinks camp life a new business. I wish he would come down this way.

The boys are all well except Ben Wells, who has been ailing several days, though not dangerously sick. Lem is detailed in QM department of the brigade. I saw an article in a Chicago paper stating that our division, or that of General Negley's, had moved from here and gone southward. This is a mistake. It is Sill's division that has gone ahead. There must be a force of at least 60 or 70,000 troops in and around Nashville.

I just copied a furlough for D. W. Holbrook [for our] com. sergt. who has leave of absence to go home. He says he will go & see my folks.

How I would like to have a sing with the young folks once more. Hoping that I shall be allowed the joy of meeting my friends ere another long year rolls by, I remain as ever your James.

Franklin Pike
[south of Nashville], Tennessee,
December 14, 1862

'Tis Sabbath eve, and I thought I could not devote my time to a better purpose than writing you a letter. How I should like to have been with you today and enjoyed your society, but how strange. Far away in an enemy's country, a soldier in the army of the U. States. But dear Jenny every day lessens the space of time which will end this unholy war; not that I repine at my lot. No one ever heard me regret, but then Jenny there are feelings that steal over one which cannot always be contented. When the quiet scenes of home & absent friends gain possession of one, he sees the many dear objects of bygone days, "then he regrets with tear and sigh; giving o'er scenes of joy gone by." 'Tis over a year; yes Jenny over one long year has passed since we parted. Many are the different places I have visited. Many the strange scenes I have passed through in the year. But never have I forgotten this, as the path of right.

I have had only one letter from you in some time, while sister has written several.

We moved camp from Nashville, the 10th day of this month, and moved five miles toward Franklin, directly south from Nashville, where [we are] encamped. Our division is the farthest advanced on this road. General Negley, a Pennsylvanian, commands our division, which consists of eight regiments of infantry and 4 batteries of artillery.[55] Generals Sill and Sheridan have advanced their forces on the roads leading to Nolensville and Lebanon, while General Van Cleve is posted on the Murfreesboro road.[56] Our line of battle is several miles in length. As far as the eye can reach, camps of men can be seen. Every piece of ground suitable for a camp is occupied. There is a probability that a battle will take place near here at no distant period. What will be the result remains to be seen. But I look with confidence. I think we have an army here that is too formidable for a defeat from that of the enemy. And a general who [is] able to command with success.

I have great confidence in Major General Rosecrans as a commander, he has shown himself one. He passes through our camp nearly every day. Day before yesterday was division review. It was a grand sight. But what was the best of all, the general said as the 11th Mich. Vols. passed him, that it was the finest regiment he had seen.

The boys have retired for the night and left me seated, penning this epistle to you. Billy Davis says I had better write shorter letters and come to bed. The night is warm and windy. I can hear the winds sweeping through the boughs of the trees. Accept my kindest wishes dear Jenny, and remember me as ever your James.

Give my best wishes to your father and mother.

> Camp Hamilton,
> Near Nashville,
> Tennessee, December 24, 1862

I received your welcome letter day before yesterday, but could not find time to answer until the present moment. There has been no change of any importance, since I last wrote you, in the Army of the Cumberland.[57] Skirmishes occur almost daily, but no general engagement. Last Sunday while foraging on the Wilson Pike south of here, we were attacked by a small party of Rebels. Quite a sharp skirmish ensued which resulted in the killing of two Rebels, the wounding of several, and the capture of six. Our regiment has gone foraging today; but few men remain in camp. News came this morning that a heavy force of the enemy were advancing on our front, but it turned out to be a scare. We are camped at present on the plantation of John Overton, said to be the richest man in the state of Tennessee.[58] He is a bitter Rebel and has given an estate valued at five millions of dollars to aid the rebellion that now exists. He has been assessed the sum of two thousand five hund. dollars to relieve the wives and children of Rebels who are suffering in the city. He thinks it a shocking cruelty, a bitter oppression that he is made to pay this amount to help keep alive the victims he has made. These are the wives and children of men whom he persuaded to enlist in the Rebel cause.

Write and let me know Jenny, what you at home think of this war. I begin to think that political aspirations is delaying the vigorous prosecution of the war, and that General Burnside was in no way to blame for his defeat at Fredericksburg.[59] If they had furnished him the needed transportation to have crossed the river when 'twas his intention, I believe now he might have been threatening the Rebel capital. He seems to have done all that possibly could have been done with the means which he was provided. He moved his army with great quickness opposite Fredericksburg, only to find his advance checked by the want of means for crossing the river. This gave the Rebels time to concentrate all their available force at this point, and when General Burnside was ready to advance, they were ready to meet him. The removal of his army across the river after this dreadful battle, [and] the skill with which he conducted the retreat, place him among the first of generals. But the intention was to achieve a great victory which would disable the rebellion, if not crush it. By some imbecile mismanagement of their furnishing him means to move, a great victory was turned into a defeat. General Burnside says in his letter to Gen. Halleck, if it had not been for unavoidable & unexpected delays which gave the enemy time to concentrate his force, we should have been certain of success.

In regard to your illness, dear Jenny, I hope 'tis nothing serious. Be very careful not to expose yourself, for good health is the greatest blessing we can enjoy.

Write soon & remember me as ever your James.

Give my best wishes to your father & mother.

––––––––––

With Bragg's Army of Tennessee weakened for the winter by detachments, and General in Chief Henry Halleck demanding action, William Rosecrans determined to strike. On the morning of December 26, 1862, more than 40,000 Federal soldiers streamed southeast from Nashville. At the head of George H. Thomas's Center Wing marched James Negley's division, including James W. King and the 11th Michigan Infantry. The Wolverines fully realized that they were advancing into their first full-scale battle.

CHAPTER 4

The Worst Scourge

December 1862–June 1863

❧

I have seen sights that would make the
blood curdle in one's veins.

Rosecrans's army slogged through rain and mud to the outskirts of Murfreesboro. December 30 passed tensely, with widespread skirmishing. At dawn on
the last day of 1862, the sleep-deprived Wolverines, who had spent a cold, wet,
anxious night on the picket line, were relieved by the 18th Ohio to pull back
and get something to eat. But their repast was unceremoniously cut short by
the rising cacophony of battle to their right. They dropped their half-eaten
meals and ran to urgent shouts of "Fall in!" As the Michiganders filed into line
of battle, James King was ordered to help remove the wagon train to the rear.[1]

Rosecrans and Bragg had each ordered a morning assault on their enemy's
right, but Bragg beat Rosecrans to the punch, and the Union flank, caught by
surprise, crumbled away in a perfect rout. Phil Sheridan's division on Negley's
right stemmed the Confederate tide, but in the process the Federal line bent
back so acutely as to form a V, with Sheridan's men comprising the left leg
and Negley's, the right. Near the apex of this V—a critically exposed point
in the Union line—the 11th Michigan was caught in the thickest of the fight,
taking a pounding but holding firm against superior numbers of Confederate
troops. Following a gallant countercharge by the Michiganders and their Illinoisan sister regiment, Sheridan's adjacent division gave way, having exhausted
its ammunition. Negley's imperiled blue line, with its right flank hanging in
the air, was ordered into a hasty retreat to escape encirclement. A desperate
fighting withdrawal ensued, punctuated with another tenacious counterattack

by the Wolverines and mercifully concluded when the artillery in Rosecrans's hastily formed reserve stymied the Rebel pursuit for good. Although the strife continued in earnest elsewhere on the battlefield, the bloody day of battle was concluded for the 11th Michigan.[2]

New Year's Day passed in relative silence around Murfreesboro, with the exception of some early morning skirmishing that sent James King back to Nashville with the supply train. On January 2 Negley's division was redeployed to bolster the Union left and arrived in time to help repel a powerful assault. The 11th Michigan took part in the battle's triumphant charge, splashing across Stones River in pursuit of the shattered Confederate ranks. The Michigan regiment's toll at Stones River escalated to 32 killed, 79 wounded, and 29 missing: 140 out of 440 engaged, more than 30 percent casualties. On January 4, in the wake of Bragg's retreat, the 11th Michigan and its brigade were the first Federal soldiers to occupy Murfreesboro.[3]

William Rosecrans initiated the arduous process of rebuilding his victorious, but thoroughly spent, army. Colonel Stoughton was appointed provost marshal of Murfreesboro, with the 11th Michigan assigned as provost guard. Braxton Bragg settled in to the south and selected Tullahoma as his headquarters. The Army of the Cumberland would pass long months at Murfreesboro before resuming offensive operations. "If the people of St. Joseph could see our regiment," James King wrote to the *Three Rivers Reporter*,

> they could form some idea of the horrors of war. When we entered Louisville, the 12th day of December, 1861, we numbered one thousand men; now we have but four hundred and sixty-six men present. 'Tis true that exposure and disease have carried away many of the victims, but the plains of Stone River has drank the blood of many of our bravest boys. Still there are enough of the 11th left to avenge their noble spirits, and with our gallant colonel to lead, we will wage a war of death and destruction to traitors wherever they may be found.[4]

Nashville, Tennessee,
January 3, 1863

Dear Jenny,

Once more I am permitted to address you.

I was in the Battle of Murfreesboro the 31st of Dec. Many of our brave boys were killed and wounded. Among the killed is [mother's nephew] Ezra Spencer, a brother of Julia's, and Anson wounded.[5] Lem is safe, was not in the battle. The night of the 30th, we were ordered to move our train to the front so as to provide the men with rations. When we reached the front we found our regiment to the right of Murfreesboro road. They had been skirmishing briskly all day though no one was injured, except Lt. Wilson of Co. F, who was struck in the head by a ball and killed instantly.[6] We slept soundly that night. But the next morning at daylight, the battle commenced. We were ordered to move our train to the rear to a place of safety. We moved the train back about two miles, where the quartermaster considered it safe. Billy Davis & I borrowed guns and went back to the field. At first we tried to find the regiment, but 'twas all in vain. By 9 o'clock the battle raged with fury all along the line. But Gen. Negley's division done the most work, and suffered the worst. The Rebels came very near flanking us on the right. In the early part of the day, Gen. Sheridan's division broke and run. It exposed our men to a crossfire, which killed many of our brave boys. All day the battle raged with fury, and darkness put an end to the engagement. I do not know the loss on either side, but it must have been numerous. Our regiment had probably 40 or 50 men killed. I found the regiment, or what was left of it, that night. Col. Stoughton had his horse shot from under him. The adjutant's horse was shot. The major, wounded and taken prisoner. Capt. Oakes was knocked down with a shell. Three of our lieutenants wounded, and 2 of them killed.[7] The train was ordered to the front. The men ate their suppers and slept on the same ground we occupied the night before. About 7 o'clock the battle commenced again. This was the 1st day of the New Year. The quartermaster was ordered to move his stores back about 6 miles. We moved in the forenoon. About 3 o'clock, [we] received orders to go to Nashville, as a body of Rebel cavalry had got in the rear and were burning wagons and supplies. We started and were near La Vergne when the train was attacked in front, and a number of the wagons burned. But our wagons were in the rear of the train.[8] We got all our stores out safe. We reached this city yesterday morning. Shall probably load the train with provisions and go back today or tomorrow.

What will be the result of the battle I cannot tell, but the 31st I did not see any advantage gained by either side. The 1st, our men drove them and gained an advantage. Yesterday the news came that Murfreesboro was taken.

I do not think our brigade was in the fight after the first day of the engagement, but was held in reserve. Excuse poor writing. I have done this in a hurry. Write soon and remember me as ever your James.

Murfreesboro, Tennessee,
January 14, 1863

Dearest Jenny,

I was just thinking what I would busy myself at this evening. I did not con-sider long. The first thought told me I owed Jenny a letter. 'Twas yesterday I received yours, dated Dec 28th. Yes Jenny more than one long year has passed, and I am still absent from all I hold dear. How much longer I may tarry no mortal can tell ... "absence makes the heart grow fonder." Could I have passed Christmas day at home, it would have been a joy beyond description. But how different ... that day we were busily preparing to move.

I was just thinking Jenny of a lesson I have often read in one of our school-books. I think you will find it in Sanders' fourth reader.[9] The caption is this. "Contrast between Peace and War." It reads thus—"Lovely art thou O Peace; and lovely are thy children, and lovely are the prints of thy footsteps in the green valleys. Blue wreaths of smoke ascend through the trees and betray the half-hidden cottage. The wares of the merchant are stored in the high-piled warehouses, the labor of each profits all." This is what it says in regard to peace.

"War"

"They have rushed through like a hurricane, like an army of locusts they have devoured the Earth. They have deluged the land with blood. The smoke rises not through the trees; the honors of the grove are fallen. The hearth of the cottages is cold; but it rises from the villages burned with fire and from warm ruins spread over the land. The groans of the wounded are in the hospitals, and by the road side, and in every thicket." Ah Jenny you at home know little of this war. You cannot have the faintest idea. I have seen sights which would make the blood curdle in one's veins. And who will be accountable at the "bar of justice" for these unholy scenes? The leaders of this unholy rebellion are alone accountable. 'Tis the vile traitors who plunged our peaceful land in civil war, the worst scourge that can befall a nation.

I passed over the battlefield [on] the 6th, where a few short days before I had seen two armies engaged in deadly conflict; when I had seen "rider and horse, friend and foe, in one red burial blent."[10] As I gazed over this field, I could see by the number of horses which lay on the ground where the most desperate carnage had been. I could see where our regiment fought on Wednesday, the day before Christmas.[11] In some places, the bushes and trees were literally riddled with balls. I have not told you yet where I am. I am seated in the upper story of a brick building in the central part of Murfreesboro, which we use as an office. The night is dark without, and the rain is gently pattering on the roof. It brings to mind a song we used to sing in school. I should have said school days; and here is a sigh for those "halcyon bygones." Perhaps you remember it. "Every tinkle on the shingles has an echo in the heart, and a thousand dreamy fancies into busy being start, and a thousand recollections weave their bright hues into wool, as I listen to this patter of the soft rain on the roof."[12] Yes Jenny a thousand fond recollections are awakened. Their bright hues come floating before me to tell of the joyous scenes that have passed. Passed into eternity. And one scene comes fresh to mind. 'Tis the night I pressed that hand of thine and bade thee farewell. Days, weeks, and months have passed since then. But 'tis not forgotten. You spoke of my not letting anyone see your letter. No one, Jenny, has ever perused a word written to me by you except myself—I do not think you silly Jenny. Why should I? Hoping this war will soon be brought to a close, our nationality once more established and peace reign forever more, I still remain as ever, your faithful James.

Murfreesboro, Tennessee,
February 8, 1863

Dear Jenny,

Sergeant Dixon was here a few moments ago to have me go and sing for a secesh lady. I declined, having urgent business to transact. I will tell you how we became acquainted with this lady. One of our lieutenants was sick and went to a private house to board. The lady was visiting there, and of course Dick and I had to sing. She said she thought us splendid singers. We were of course astonished at this, as we had never been called so before. I think there must have some flattering about her words. Don't you, Jenny? She wanted to know if our sweethearts were good singers. I told her I had never heard mine sing. But I guessed she was. She questioned us about the North, whether the people of quality, as she called them, worked. How we got along without servants. She thought we were people of quality, at least we appeared as such. I told her that our people were not only people of quality but qualified for anything. That is, they could labor, or appear in a drawing room. She finally finished by telling us that she had a very charming daughter Alice, whom she thought would like the ways of the North much better than the South. And she intended selling her property and going North. Dick advised her too [that] Alice might learn many valuable lessons.

Our men have skirmishes nearly every day, with the enemy coming out beat in almost every instance. Today, thirty of the Rebels were brought to Col. Stoughton, who is provost marshal. The courthouse is right opposite our office, and we see all the Rebels that is brought in. Hardly a day passes but what a squad of these men are brought in. They are the worst looking set I ever saw. They look as though they had got all the Southern rights they ever asked for, and were now willing to go home. There is a Rebel hospital, the next room to ours. They have as good care as our men. The government provides them rations and medicines, and the women in the town bring them many delicacies. I have talked with many of them who say they were forced to go into the army or be hung, and of course they went in. They tell about the suffering of the South. That is, the people. Men who never knew what it was to want, are now suffering for the necessaries of life. Civil war in its mildest form is a horror. Jenny you have not the least idea of the misery and destruction which this war has caused. No one can put the slightest estimate, unless he has witnessed. To pass through this country where a few short years, nay even months [ago], all things were beautiful, where plenty and prosperity reigned. But now, nothing but destruction meets the eye. As we pass along the leaving roads, we see no fences. A little back from the highway, we see the ruins of a large brick building. This house was once a peaceful home. Here a father sat and enjoyed his home, which he had labored to gain, as mother watches over the interests of her children. Where are they now . . . they are gone. The monster secession entered the sacred circle. The male members of the family have turned traitors to their country. The mother and the daughters have been sent south. The home is left to the mercy of the Northern soldier who, to avenge himself against some brother's wrongs, applies the torch, and the consuming flames speedily do their work. That once splendid home is now a mass of ruins. Ah Jenny there is an awful penalty resting on the heads of the leaders of this rebellion. May each of them meet a traitor's doom.

'Tis getting late, and I must bring this to a close. Give my best wishes to your father and mother. Goodnight. Your James.

––––––––––

Across the country, there was cause for both hope and concern. Ulysses S. Grant's drive on the critical Mississippi River stronghold of Vicksburg alternately advanced and sputtered. Out east, the demoralized Army of the Potomac changed commanding generals again; its latest commander, Joseph Hooker, began preparations to march an enormous, bountifully-equipped host against Richmond. But with the bloody conflict dragging on, and with controversy stirred up by both the Emancipation Proclamation and the draft, the antiwar faction of the North, known as the Copperheads, was growing bolder and gaining influence.

Jenny's family, meanwhile, experienced conscription firsthand. Michigan

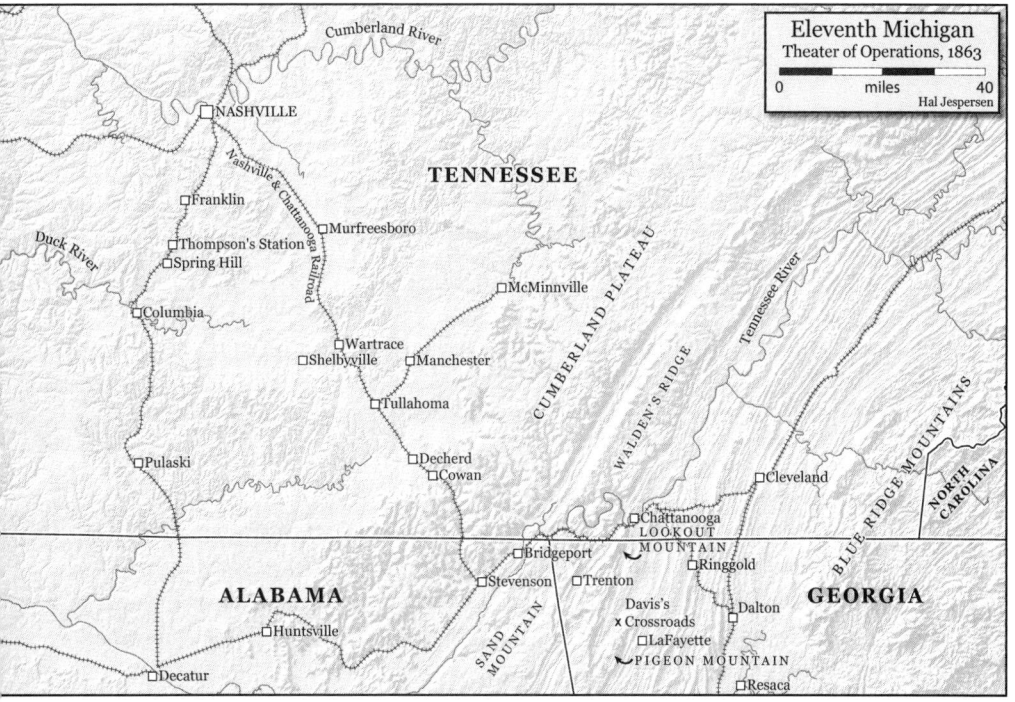

implemented a draft in February 1863 in response to the state's failure to meet its federal enlistment quota. Jenny's father, forty-two-year-old blacksmith Darius Ambrose Babcock, was drafted on February 11. Jenny's uncle, John Wesley Blodgett, traveled to Detroit and signed an affidavit testifying that Babcock had suffered a debilitating head injury nine years before, implying that he was unfit for military service. Evidently this assertion of disability came to naught, for Babcock ultimately paid a substitute by the name of Louis Gerichten to fill his place in the army.[13]

James King was promoted to quartermaster sergeant. In this role, he was accountable for taking charge of regimental property, and for directing soldiers detailed for quartermaster duty. As quartermaster sergeant, James also supervised teamsters and laborers and oversaw the wagon master and regimental train.[14] This new rank assured him a level of comfort and safety above that of most soldiers, something he had already enjoyed through much of his enlistment, thanks to frequent assignments to hospital, clerk, and quartermaster details—noncombatant jobs that often provided a roof over one's head, and which soldiers referred to as "bombproof." James had previously commented to Jenny that he appreciated such comforts, yet his actions at Stones River, joining in battle outside his line of duty, proved that he had no desire to avoid combat.

———

Murfreesboro, Tennessee,
February 28, 1863

Dear Jenny,

I should have written you ere this, but have had a very lame hand. This is the first I have attempted writing in a long time. My hand is yet so lame, I can hardly write. Therefore you must excuse me if I do not write a long letter.

I received a letter from you last week. Also one from sister, stating that your father had been drafted. Write soon and tell me all. Your ever faithful James.

Murfreesboro, Tennessee,
March 4, 1863

The last time I wrote you my hand was very lame, and consequently my letter rather brief. But that was only a few days ago, and you must give me credit for writing so soon again. The last letter I received from you was written the 22nd of last month. I also received one from sister the same day. You spoke of singing school, and she also wrote that she was getting ready to go. How I should like to have been with you and taken part in the happy exercise, but I feel confident that the day is fast hastening forward when I shall again be permitted to share the joys of home and friends.

We are looking anxiously forward for the taking of Vicksburg. That place in our possession, the noble Mississippi once more free to navigation, it will be an important step toward bringing the war to a close. I think General Grant competent for the great work, and will use every means in his power to accomplish the undertaking.[15] If the Rebels evacuate Vicksburg and escape Grant, they will reinforce General Bragg, then look out for warm work in this direction. But there is one consolation. We have a leader who never knew defeat; and here, right here, the drama of Stones River may again be enacted. If this must be done, let it come, the sooner the better. The prospect looks more cheering Jenny than it ever did before. All there is to cloud or shadow is the sympathy shown these black-hearted traitors by a few, no I will not say a few. But by some of the Northern peace men.[16] While we are here undergoing the dangers and privations of a soldier's life and attempting to put down the vile miscreants who are aiming at the lifeblood of our beloved country! [With] many of our brave men pouring out their life's blood in the cause of the Union, there is a party at home who sympathize with the traitors and are doing all in their power to clog the wheels of government and stop the prosecution of the war. They find fault with the proclamation of the president and in every act of the administration. The Detroit Free Press says if the president had not issued the proclamation there would have been volunteers enough to fill all demand. Now Jenny I do not think this issuing of the proclamation has made the difference of one man. Why don't they find fault with the Southern Confederacy who have trampled all rights and liberties under foot? Oh no, not a word to say, they are a much abused people and the North was the aggressors. I cannot write any more at present. Give my respects to your father and mother. I am as ever your James.

Murfreesboro, Tennessee,
March 13, 1863

I received your letter dated March 1st and now seat myself to reply. In it you said you hardly knew what my thoughts were. Jenny knew my thoughts once, they have never changed. You say you have not received many letters from me of late. I have written you two letters before I received your last, explaining the reason of my not writing, and last of all, you ask me if I am tired of writing you. There is no one that I converse with in the language of the pen that gives me greater pleasure than your own dear self, Jenny. Now I think when you consider everything, you will not blame me so much after all.

The Army of the Cumberland remains nearly the same as when I last wrote you, except the Franklin disaster which you must have heard of ere this.[17] I received a letter from brother John yesterday; he is at Franklin. He also enclosed a letter from his wife. I think I showed you one of her letters before I left White Pigeon. She shows the same spirit she did then.[18] A few nights ago we had the pleasure of listening to an address delivered in the courthouse at this place by the celebrated orator and poet, T. Buchanan Read.[19] The room was in the upper story of the courthouse, and was filled with officers and men to overflowing. Many could not get in. On the walls of the room were the words: Ft. Donelson, Pensacola, Shiloh, Hartsville, and many other names of the bloody fields which were purchased with the dearest lifeblood of the nation.[20] These were surrounded by wreaths of evergreens. The 15th Regulars band was in attendance and played many beautiful airs. General Garfield, chief of staff for General Rosecrans, came and addressed the audience in a brief but appropriate manner.[21] It was expected that General Rosecrans would be present, but he was not. General Garfield said he had come to fill his place, though he was afraid it would be poorly filled. He said this was a peculiar kind of meeting; but the Army of the Cumberland would never flinch from its duty, even if the whole North turned against us. There were peace men in the North who were doing all in their power to aid the Rebels. But he was satisfied that the masses at home were with us. He then introduced the speaker, who said he did not come before us as an orator, but as the bearer of great joys and glad tidings. He had been at the North, and had witnessed scenes which made him feel certain that the army would be sustained by those at home. There were many peace men at the North, but their days were numbered. He read the resolutions passed at the great Union meeting at Cincinnati. These were received with cheers, but when the name of General Rosecrans was mentioned, 'twas no cheer, but a prolonged shout. I think him the ablest of American generals. The speaker read several poems portraying the deeds of the men of '76. And to end the scene, repeated a poem entitled "The Oath," written out of the bitterness of his own heart.[22] He had lost a brother and two nephews who had fallen sacrifices since this terrible war began. It was truly thrilling and beautiful. You spoke of an exhibition in your uncle's district. The mention of such scenes recalls old times. You also mentioned Miss Sutton. Is she in Michigan? I think sister wrote me that she went to New York. If she teaches this summer on the prairie, are you going to attend her school? My space is getting limited, and [I] must close by asking you to write soon. Yours as ever, James.

Give my best wishes to your father & mother.

Murfreesboro, Tennessee,
March 23, 1863

Dear Jenny,

As I seat myself to acknowledge your kind letter, I cannot express the joy it gave to receive it. I had not had any letters in several days, but on going to the office yesterday evening I found four letters addressed to me, and one of them in your well-known hand. You may think I have a very extensive correspondence to receive four letters in one day. One of them was from an aunt in New York, one from my brother at Franklin, and the other from brother Henry.[23] In his letter, he said he saw you in Three Rivers a few days before his writing. I hope your father will not be compelled to go. It would be cruel, when there are so many who could go as well as not. The young man who will stay at home in this, the time of his country's danger and peril, has not the least spark of patriotism and is unfit to be called a man. There are many in our own town Jenny, perhaps you know who I have referenced. I except none.[24]

There are two persons in the room beside myself. One of them is the sutler of the regiment, whose name is Moore. He is a splendid fellow and boards with us, the other is the quartermaster's nephew whose name is [Lindley R.] Harkness.[25] He is a very good fellow. He says he has a sweetheart, and to hear him talk, you would think her equal could not be found on the face of the Earth. Billy Davis has just come in. He stays with us. Billy and Harkness have set the sutler's whiskers on fire with his cigar. But how do I know that this will interest you? I will change the subject. How I should like to be with you this evening. I am sure it would be far pleasanter than conversing with you on paper, and I consider this no disagreeable task. I cannot write any more at present. Write often Jenny, and long letters. James.

Give my best wishes to your father & mother.

Day before yesterday, a very sharp skirmish took place in front; a sharp cannonade was kept up quite a while. But the artillery became too hot for the Rebels and they withdrew. The prospects for an early close of the war look very bright. Grant must clear the Mississippi. Rosecrans, whip them in Tennessee, and Hooker get possession of Richmond, then secession will be on its last legs. With such generals to lead as Rosecrans, Rousseau, Negley, Crittenden, and many others, we cannot help but be victorious.[26] And there is one, who watches over the destinies of nations, in whom we put our trust.

[From James to his uncle, the educator and author Anson De Puy Van Buren—ed.]
Camp, 11th Mich. Vols. Inf.

Murfreesboro, Tennessee,
March 30, 1863

Dear Uncle,

There are a few government buggies in the streets today. I suppose you know what I mean by a government buggy. Let me describe one. They are nearly the same pattern our farmers use at home, though much stouter and stronger built. The box has 6 loops built over it and over these loops, a covering of canvass cloth. The locomotive properties of the vehicle is neither steam or horse power, but centered in 6 mules which are all driven with one line, the driver riding the off-wheel mule. Often the driver is a contraband who belonged to some secesh who waked up some morning and found himself minus a darkey.[27] Probably the darkey had heard of a proclamation issued by one Uncle Abraham who was high in authority, declaring all slaves in the land of seceshia free henceforward & forever, and had taken a notion that it meant him. The Negroes come here in flocks daily, and are used about the hospitals and government works. They are all colors and sizes. From the white to the coal black. A freckled Negro and red hair is no uncommon sight. But I supposed you have seen enough of niggerology to have witnessed all these sights without my telling you. But by the way, our cook is a contraband who has been all over the South, and yesterday he gave me a very good description of your home in Mississippi. He used to run on a steamer from New Orleans to Yazoo City.[28] He says they were laid up a week at Willow Dale, 6 miles below Yazoo City. He is the most intelligent darkey I ever saw. Has seen all the big guns of the Southern Confederacy, from Jeff Davis down to the guerrilla Morgan. He was at Shiloh and Stone River. But before I go further, I will drop the Negro subject and talk about something else. Though I am certain that this is a very important subject with many at the North. Among them the editor of the Free Press and Chicago Times, a disease which they have contracted since the issuing of Uncle Abraham's proclamation.

You say spring will soon return with its birds and blossoms, but the winter of our national discontent is still with us. I tell you uncle, in spite of the grumblings at home and the efforts of the traitors in the South, the dark clouds of civil war which now hang so threatening over our beloved country, ready to descend and destroy all good, must raise and give place to a sweet peace ere the autumn reaches us. Vicksburg must be ours. Bragg must be driven from the borders of Tennessee. Hooker must assail the Rebel capital and gain possession of that hotbed of treason. The starting point of secession must have the stars and stripes waving peacefully and free from her steeples and forts. When I say the hotbed of treason, I allude to the city of Charleston, South Carolina. All of this can be done ere the winter reaches us, and it must be. Our armies never were in better fighting condition than they are today, and backed up by the good cause they are fighting for, will muscle to certain victory. The coming struggle is one of unequaled magnitude, and will be fierce and bloody. But we all know the great responsibility resting upon us, and are equal to the task.

In your letter, you said that Ezra's remains were to be sent home. That was the intention, but he had lain so long it was impossible.[29] Anson was sent to Louisville and I expect will be discharged.

When you write the cousins in the army, give them my best wishes.

Every preparation is going on for an early move of the army. Your affectionate nephew, Jas. W. King.

Murfreesboro, Tennessee,
March 31, 1863

Dear Jenny,

I am glad to learn that your father was not compelled to go. 'Tis cruel to take men from their families when there are so many young and single men who could go, as well as not. I saw a letter from your Uncle John [John Wesley Blodgett] today, written to Lem. He gave a doleful picture of a soldier's life. It seemed that he went to Detroit when your father was there. I think from his description, soldiering must be far worse in Michigan than it is in this land of seceshia.

I received a letter from sister yesterday, and also yours. By the tone of her letter, I concluded that there must have been some trouble between you and her, though she did not mention it. If so, what is the cause? Do not mention this to anyone. I trust you Jenny to keep it a secret. She did not say as you had had any trouble. She may have written in the way she did to joke me. Remember me to your father & mother. James.

I wrote this last evening. Today is the 1st of April, known as All Fools' Day, and many is the joke that has been perpetrated by the boys.

Murfreesboro, Tennessee,
April 7, 1863

Windy march has left us, and April is a welcome visitor with her many smiles; her sunshine and showers. The roads are in good condition and everything seems in readiness to make a forward move. Yet we may not leave here for many weeks. If Bragg is receiving the reinforcements they say he is, our general may await his coming and be glad to receive him behind our own fortifications. They could not whip us, Jenny, with five times our force. Let us fight them from the position we now occupy. But I hardly think they will venture back here again, unless they receive overwhelming reinforcements from Richmond and Vicksburg, and I hardly think they can spare them from either of those places, as they have all they can attend to.

Madam Rumor said yesterday that Charleston was in our possession, and the glorious stars and stripes floated over the hotbed of treason. But 'tis hardly credible.[30] I would to God it were so. The traitors boasted of this as the strongest of their strongholds, and was invincible to our army and ironclads. We have heard nothing new from Vicksburg in several days. I hope Grant may be successful in his work. I am confident that ere the summer endeth, this unholy war will be ended. I have talked with many of the Rebels, and they say the Confederacy cannot support its armies another year on account of the great scarcity of provisions. Another three months will bring great changes in our national affairs, and I feel certain it will all be for the best.

Yesterday was a grand gala day for the regiment. A few evenings ago our boys met and subscribed six hund. dollars to purchase a sword to present to Col. Stoughton. The preparations were all made, the col. in the meantime not knowing anything about the proceedings. Col. Stoughton is provost marshal, and does not have command of the regiment. Capt. B. G. Bennett, who has command of the regt., formed it and marched it to the parade grounds. A large crowd from the different regiments had assembled to witness the scene, among them General Negley and Col. Stanley, who commands our brigade.[31] Everything being in readiness, the col. was sent for to come and drill the regt. You cannot imagine his surprise when he saw what was going on. Captain Mudge of Co. B presented the gift after making a few appropriate remarks.[32] The col. responded in language which came from the heart. It was a magnificent present. But 'twas not the intrinsic value which made the scene interesting. There was a feeling expressed on the countenances of all which cannot be described. Nineteen long months, months of darkness and danger, have we been with him, always at our head and never from his post in time of danger. His dangers have been ours. Our privations his privations, our perils his perils. He deserved the gift, and when the storm of war is o'er, may he lay it aside and keep it sacred in remembrance of the men of the gallant 11th, who gave it as a token of the high esteem [in] which he was held by them.[33]

Lem has just come into the room. I sometimes think he is homesick. He thinks if he had it to do over again, he would not enlist. I cannot say as I agree with him. I have found no worse fare or greater hardships than I expected on the start. I cannot write any more at present. Remember me to your father and mother. Don't fail to write soon. James.

Murfreesboro, Tennessee,
April 17, 1863

Dear Jenny,

Your letter of the 6th instant is at hand. You wished me to reply as soon as I received it. I took time to read it over twice, Jenny, before doing your bidding, and now I am seated to reply. I had looked several days in vain for a letter from you, but were disappointed until today. I forgive the delay, Jenny, on account of the length of your letter. Such letters are indeed treasures. Never think of wearying my patience with too long ones, those are the kind I like to receive, and especially from Jenny.

I do not know as I can agree with you in regard to persons at home enjoying themselves. I do not wish my friends to hang their heads and cast a gloom over everything and everybody because I am a soldier and forced to endure the dangers and privations of such a life. The fact is, Jenny, there is nothing like cheerfulness, under any circumstances. We must make the best of everything, come good or bad. How I should enjoy a ride on the lake with you, if only permitted. You must have had a pleasant sail and joyous time the day Hattie visited you; I remember once upon a time 2 as jolly loads as ever filled vehicles visited a school one pleasant summer day.[34] Though I have often thought since, 'twas more to see the school miss than the school. At least that was the case with me, though I would not own it then. The school dismissed at noon, and hurrah, a sail on the lake. Did we not have a glorious time? That was no soldiering, indeed. Such scenes as these, Jenny, never will pass from my mind as long as thoughts remain. You spoke of the death of Mr. Goodsill.[35] Even among their friends people will die, as well as in the army. But not so often as here, not a day passes without seeing a fellow soldier borne to his long home. 'Tis a sad, sad sight to behold the manly form laid in the cold, damp grave, but such is life. "In the midst of life we are in death."

The election returns I had heard by way of the St. Joseph County papers. You spoke of Copperheads having the rule. Men who have opposed the prosecution of this war will be sunk deeper in degradation in after years than the Tories of the Revolution.[36] Their days are numbered. I had not heard as sister intended teaching this summer. What school did she have in view? I do not think she would enjoy herself as a teacher. Billy Davis tells me to write a few lines for him. He does not know who I am writing to, but that is his way. I have just lit a candle to see to finish my letter. Twilight has hardly faded from view. The western sky is lined in yellow golden clouds.

We are looking anxiously for further news from Charleston. It seems that the ironclads met with serious impediments in trying to pass beyond Fort Sumter. I do not think the struggle is ended in them trying to take the city. If those chains and torpedoes can be removed, the city is at our mercy. It is from beyond a doubt that the ironclads can pass the forts.[37]

The boys are all well, except Ben Wells who has been very sick with fever; in fact is pretty sick yet, but much better than he has been. I have written to his friends twice in regard to him. I think if he gets able to be around, he will be discharged.

I had a letter from brother John, written while at Franklin. He wrote the day before Van Dorn attacked General Granger at that place.[38] I am anxious to hear from him again. There was a train of cars captured on the railroad between here and Nashville last week. There were two officers of our regt. on board who were captured. The guerrillas marched them in the direction of Lebanon about 18 miles, and then paroled them.[39] They reached camp Sunday. I cannot write more now. Give my respects to your father & mother, and accept this from James.

Murfreesboro, Tennessee,
April 23, 1863

Dearest Jenny,

Your letter of April 12th is at hand. The style of your letter, the force and openheartedness with which you penned it, gave me great pleasure. Rebecca did not tell me there was any trouble existing between her and you, and now since the receipt of yours, I am confident that all that was said was in jest, and meant for me more than anyone else. No Jenny, friendship is not a mere name with Rebecca. She is a good, noble, and generous sister, well worthy of your deepest friendship. If sister had said anything to wrong you, I should tell you. But on the contrary, she did not. She has always, when writing to me of you, extolled you. Rebecca wrote me that some of her lady friends had been giving her advice in regard to going with young men, and she thought some had better take the advice to themselves, or should have done so about a year and a half ago. I suppose she was thinking that I did not spend many evenings at home, and mentioned this by way of remembrance. Never mention what I have written to her, and bury all I said in the other letter in the past.

There has been a grand review of General Negley's division today. To see so many troops all in uniform is a grand sight. There is a great movement of troops at present in this vicinity. Crittenden's corps is moving on McMinnville, and McCook's on Shelbyville.[40] Our corps is awaiting orders. You must not be surprised to hear of another bloody battle in Tennessee.

You said Billy Davis had written your father and pretended to know everything. Billy knows nothing of my affairs, but does some close questioning.

Billy Bournes is well, and the same comical personage he always was.[41] I saw Lem a few hours ago. He had been on review. He is division train master.

You said some think peace near at hand. I think it is myself, but then we may be mistaken. The future will tell us. If the fleet had been successful at Charleston, it would have been a grand step in the right direction. I do not think our men have given it up as a bad job. A month may work wonders in the affairs of our nation; and may it be in the right way. Hoping that an honorable peace may soon be established, and that I may be permitted to return to friends who anxiously await me, I remain as ever your faithful James.

Murfreesboro, Tennessee,
May 4, 1863

Dear Jenny,

'Tis a splendid morning, a little too warm for comfort, but what care we if it is a little too warm when one is greeted by the sight of a thousand flowers?

The roses, snowballs, & pinks are with us, and a thousand other of their kin bear them company. Nature has created wonders since the 1st of January in this city of Murfreesboro. The streets that were then so cold and forbidding are now clothed in a yard of the richest green. Let me notice the view one gets in passing through the streets of Murfreesboro. I think you at home would call it a military view. The first thing of interest is the courthouse, a splendid building [that] would do honor to any county. As you gaze at the entrance ways, you see officers and soldiers going in and coming out. In the yard is a stack of guns belonging to the guard. One of the soldiers is on guard, guarding the arms. Several cannon stand in one corner of the courthouse yard. In the streets are government vehicles drawn by 6 mules and driven by some of Uncle Sam's boys, or an intelligent contraband. Every now and then, an officer mounted on a superb charger, putting on as many airs as Julius Caesar or Alexander the Great, dashes down one street and up another at a killing pace. And another interesting feature in the program is the females who appear in the streets. They are generally Northern ladies who have accompanied their lords into this land of seceshia. Hardly an hour passes without seeing some of these ladies with their lords pass by on horse. Many are magnificent riders. Mrs. General Rosecrans was here several days. They passed our quarters several times, she riding a small chestnut-colored horse.

A regiment of cavalry is passing. Going on a search. We have just received orders to keep five days' rations on hand and be ready to move at a moment's notice.

My brother is still at Franklin. It was a few days since Billy Davis received a letter from your father, a few days since, at least, he said it was from him. Lem and the other boys are well, except Benjamin Wells. He does not improve very fast.

No more this time. Write soon. James.

Murfreesboro, Tennessee,

May 12, 1863

I have been to a wedding today, what do you think of that Jenny? I will tell you all about it. Early in the day Lt. Whallon came to me and says, "Jim there is to be a wedding today on the battlefield, and I will procure passes if you will furnish horses."[42] We were soon ready, and started for the scene of action. Of course we could not let so novel a sight pass without witnessing it. Judge of our disappointment when we found we were too late to witness the interesting cere-mony. But what was my surprise to find the bridegroom to be an old acquaintance whom I had not seen for over a year. They were strolling arm in arm along the bank of the river, in the shade of the forest trees which grew there. He saw me and hailed me, gave me an introduction to the bride. I wished them a long life

Negley's division charges across Stones River. (*Frank Leslie's Illustrated History of the Civil War*)

of happiness and passed on. This is rather romantic Jenny, but I think much more romance than sense. But as the old saying is, "everyone to their notion." Some of the party wished to visit the part of the field where Negley's division fought, and where the 11th covered themselves with glory. It is to the right of the railroad running to Nashville. Here in the cedars and a small open field they fought. The graves of our fallen comrades remain to mark the spot. The wild flowers bloom around, and the sheltering trees lend their grateful shade o'er our comrades who fell a sacrifice in that awful contest. Among them is Ezra Spencer, the bereaved of all who knew his generous spirit. To the west and north of this is a cleared field, and in this field is a brick kiln where the 19th Illinois charged bayonet on the 30th of January and drove the Rebel skirmishers from its protection.[43] It was near this spot where the brave General Sill was killed while bravely leading his men against the enemy.[44] A plantation house stands a little back of this. We repaired to this house and were invited inside by a fair Rebel. There were three ladies in the room besides the one spoken of first, and [the] man of the house. One of the fair ones pretended to be Union, the rest secesh of course. A cannonball from one of our batteries had unceremoniously entered

"The wounded piano": a nineteenth century postcard image of the instrument owned by Giles and Mary Harding of Murfreesboro. Damaged by a Federal cannonball, this piano became something of a tourist attraction. (Author's collection)

near the door, broke and shivered the leg of a piano that stood in the room, and passed out the side of the house. After a little coaxing and a great many excuses on her part, she finally committed to play a tune on the wounded piano. She played several pieces, among them the "Manassas Wagon." This was secesh in the strongest terms. This caused some remarks. She said "that was a great victory for the Confederacy and a deep wound to us." Lt. Hall of our regiment made one of the party. He was taken prisoner and tried the sweets of Libby Prison. He replied, "It was a deep wound, but an ointment called Fort Donelson, Shiloh, and New Orleans had more than healed the wound." She concluded there was no use trying to get ahead of Yankees talking, and we took our leave.[45] On our return, we crossed the river where Negley's division made their grand charge on the 2nd of January. I saw an illustration in Frank Lillies [Frank Leslie's].[46] It is a true picture. There is nothing here to remind one where the conflict raged so dreadful, only the battered appearance of the trees and the many graves which dot the plains. My space is limited, and I must draw to a close. Write soon, and in the meantime remember me as ever your James.

Murfreesboro, Tennessee,
May 21, 1863

Dear Jenny,

The army has orders to be prepared to move forward at a moment's notice. Our surplus baggage has all been sent to Nashville, and everything [is] in readiness for a journey down into Dixie. There is a rumor that our regiment is to be detained as provost guards. How true this is remains to be seen. The col. does not think we will move.

The news from Gen. Grant is extremely cheering. If Grant can hold Jackson, the state capital of Mississippi, the army of Vicksburg will have to starve or surrender. This is their only means of supplies, and that is now in the hands of Grant.[47] I do not look upon the crossing of the Rappahannock as a reverse to the Federal arms. No earthly power could have saved Lee's army from total destruction. But the heavens opened; the rains deluged the earth, [and] the angry and boiling river rose, threatening to carry away the pontoon bridges, their only way of getting supplies. If [Hooker's] communication had been interrupted, he must have surrendered himself and [the] whole army! But like a prudent general, he re-crossed the river. I am confident, Jenny, had it not been for that storm, the glorious old flag would have floated o'er the walls of Richmond ere this.[48]

In the streets are many of Uncle Samuel's vehicles, loading with quartermaster stores. Everything is being transferred behind the fortifications. I cannot write more at present. Give my best wishes to your father & mother. James. Benj. Wells has nearly recovered.

May 22, no orders to move yet, cavalry has been moving all night.

Murfreesboro, Tennessee,
May 23, 1863

Dear Jenny,

When last I wrote, we had orders to be ready to move at a moment's notice. We have waited anxiously for the word "forward," but it has not been given. I said anxiously. Yes Jenny we have lain so long idle, we wish for something to break the dull monotony which is now upon us. But our leader knows when to move, and there is no use of being impatient. When the time comes to strike he will, and it will surely be in earnest. The news from General Grant is truly cheering. Haines' Bluff is ours, captured 57 pieces of artillery, a large number of prisoners and the outer works of Vicksburg captured by our forces.[49] This is glorious news if true, and it comes from a source not to be doubted. If Grant gains a great victory, in all probability the fragments of the Rebel army that escapes will join Bragg and attempt to crush Rosecrans and drive him from Tennessee. This is their only salvation, and if this be so, our army may quietly lie here and await the coming shock.

I am seated at my desk in a pleasant corner of a front room that faces the street. Across the street, in front of the courthouse, are several horses tied. Billy Davis & Holbrook have just come in and have a lot of nuts and raisins to eat. They appeal to me to help them. I cannot resist.

Jenny you may sometimes think my letters are cold and uninteresting, that my feelings have changed; but 'tis not so. Often when I write you, I do not feel in the mood to write. But could you see my heart, you would have no fears. There is one my heart beats fondly for, and that is thy own dear self.

In one of your letters, you asked me why women could not share the dangers and perils of war. I will tell you Jenny, I look upon woman as an embodiment of peace and gentleness. She is not fitted to pass through the trying scares which war always brings. I think you can see my argument. James.

Murfreesboro, Tennessee,
June 5, 1863

Dear Jenny,

'Tis a dark and stormy night. The rain is falling in copious showers. All is dark and cheerless without, but comfortable inside the room. The order has not come to march yet, and there is no telling where we will move. I hope it will be soon.

There has been some excitement today. One William A. Selkirk was hung today for the murder of a citizen by the name of Weaver. I will not give you a description of the scene. Such things are too gloomy to relate. I copied his confession day before yesterday. It was sent to headquarters and filed.[50]

I received your letter, written the 23d instant of last month but postmarked 29th. Joseph did not set the example you supposed he did.[51] She was a Northern lady. Such being the case, I suppose the volunteer service will not be as large as you supposed it would in case he had taken a Southern beauty. If Joseph had set the example, I do not think it would have become contagious. I have seen many splendid women in the South, but they cannot [illegible] in with our Northern girls. That is my opinion. Others might not think so. The ladies, these wondrous beauties of which you speak, have one very bad habit, and that is nearly all use tobacco & snuff to excess. What would a woman in our civilized land be thought of with a great quid of tobacco in their mouth, I leave you Jenny to imagine the picture.[52]

I know many such persons as you speak of in regard to enlisting, but do not know who you can have reference to in particular.

The band is playing for roll call. I can hear Co. J's orderly as he calls the names of his men.

The quartermaster and sutler are conversing about home scenes. Billy Davis is making the bed. You asked me in regard to Lem. He thinks this is a good school of experience for him. I do not think that a soldier's life has affected or demoralized him in the least.

There has been considerable skirmishing of late. Sharp cannonading was heard at Franklin nearly all day yesterday. I am anxious to hear the particulars.[53] I had a letter from brother yesterday. He was well. They were still in Franklin.

The news still continues cheering from Grant. It seems that the South Carolina army are operating on the Mississippi.[54] There is no doubt but what our forces will be successful.

You spoke of learning to farm now that the men are all in danger of being drafted. You would make a pretty looking farmer Jenny. Tell me how you progress in your new avocation. Remember me to your father & mother. James.

Murfreesboro, Tennessee,
June 13, 1863

Dear Jenny,

I have just returned from a ride in the country. The commissary [Commissary Sergeant Lindley Harkness] and I saddled our horses and rode to where the green fields and farmers' houses meet our view. Yes Jenny, though civil war has swept in all its homes over the length and breadth of Tennessee, 'tis still beautiful in its isolation. We took the Liberty Pike and went in a southeasterly direction until we came to the pickets. There, we cut across through the woods to the Woodbury Pike. This country reminds me much of that in Florence between Three Rivers & White Pigeon. Only their groves are much finer, and one meets with many trees and shrubs which do not flourish in a more northern clime. [In] many places near the fortifications, these groves have all been felled by the ruthless axe to give range to our artillery. One has many sorrowful thoughts as he gazes where destruction has swept the lovely plains, but then you think, 'tis a curse brought on by themselves. Wonderful has been the change since we first entered the place. One long year has rolled away since the time of which I speak. All the houses were standing, surrounded by their numerous Negro cabins, the master taking his comfort and the Negro tilling the soil. But a change has come. The Southern army gained possession of the country, and then in turn we are back again. Now the fine buildings nearly all stand tenantless, and the

Negro has taken advantage of Uncle Abraham's proclamation. Many of the inhabitants who professed Union sentiments in the strongest terms, and had even went so far as to take the oath of allegiance, are not to be found. They have decamped to the regions of the sunny South. Yes, but Jenny I commenced telling you about my ride. I believe we left at the Woodbury Pike. If you follow us Jenny, you must go across lots with us. Through the woods and cleared fields we go. Here we come to the Manchester Pike, cross it, and the next road we come to is the Shelbyville Road. You may think there is considerable cross lot traveling. But I learned this when small. Who would not go half a mile farther for the sake of going cross lots than in the road? Down this Shelbyville Road, the distance of 30 miles, is the town of Shelbyville, where the Rebel Bragg is encamped. We returned to the city by this road.

I received your letter written last Sabbath today. Anxiously do I look for the arrival of these missives. "They are bright spots on life's pages." Short letters are welcome Jenny, but long ones are preferable. In your next, tell me who teaches your school and what branches you are studying. Where is Milo Hammond?[55] I have not heard from him since last summer.

June 14th. My brother's regiment have moved from Franklin and gone to Triune, a small place between here and Franklin. They left the day before the battle. He has been quite unwell but was some better when he wrote. I am in hopes they will be sent here.

My best regards to your father & mother. The boys are all well. Ben Wells has gone to the convalescent hospital, not being able to march in case we should get orders. James.

Murfreesboro, Tennessee,
June 21, 1863

Dear Jenny,

I have received a letter from you, written the 8th of this month. In yours, you spoke of your cousin Ezra [Fulkerson] and desponded of ever seeing him again.[56] You thought if he were in the army, you would have more encouragement. Seems to me Jenny, I remember something in regard to the army, especially the soldier. It was about the time that the cry for volunteers was raised throughout the land, and some deemed it the same as burying their friends to have them go into the army. Many advised their friends not to enlist. They were, many of them, never to return, but none sorry that they had followed duty's call. Nearly two long years have rolled away, and we now number in the past the time of which I speak. As I review that lapse of time, and think of the many scenes which has attended us through our journeying, it seems all a dream. It has been no pathway strewn with roses, but in many places rugged and rocky, and filled with thorns. Yes Jenny many have been the hardships which we were called on to undergo, many the dark cloud that hung threatening o'er us, but we have escaped and passed through them all. To look back, we see among the dark clouds some bright spots and gleams of sunshine that will remain fresh in memory as long as reason holds her sway. Among them is the remembrance of the many cheering words which have reached us from the loved ones we have left behind. How dull must life be without friends to cheer us along life's wayward road. Friends, Jenny, are like the lighthouses on the bold and rocky shores of ocean that lights the mariner safely into port. Temptations often surround. But the thought of dear ones praying and watching for our safety tell us to keep the straight and narrow path of right.

We are looking anxiously for the fall of Vicksburg, and awaiting news from the Potomac.[57]

The boys are all well. Lem and I took a good long ride in the country the other day, which I would describe, had I time. Remember me to your father & mother. James.
You spoke of visiting often at my home. I am sure Jenny you are always a welcome visitor.

Murfreesboro, Tennessee,
June 23, 1863

Dear Jenny,

Orders have been received to march tomorrow morning. In what direction, I am unable to state. Our goods are all packed, and preparations made for a long march. The first opportunity that offers, I will write again. Remember me to your father and mother. May the good angels guard and protect you is the earnest prayer of James.
June 24th, the army is moving south.

———

After spending almost six months at Murfreesboro, William Rosecrans finally advanced the Army of the Cumberland toward the strategic objective of Chattanooga, a vital rail link between the eastern and western halves of the Confederacy, setting into motion one of the most dramatic sequences of events in the entire war.

CHAPTER 5

This Cannot Be a Defeat

June 1863–April 1864

I . . . witnessed for over three years the brave acts of the bravest soldiers that ever
fought on [a] battle-field, and never, during all that time, did I witness a braver act
or a more heroic deed than was performed by Comrade King in the voluntary part
taken by him in the charge of Mission Ridge.
—*Captain Borden Mills Hicks*

The bulk of Bragg's army was drawn up in the vicinity of Shelbyville and
Wartrace, with headquarters at Tullahoma. The rugged terrain of the region
made for hard marching and offered the Confederates opportunities to check
the Federals at the mountain gaps. But the Union army promptly seized key
passes, including Hoover's Gap, enabling Rosecrans to turn Bragg out of his
position. The unexpected swiftness of the Federal advance even raised the
tantalizing possibility that the Rebel line of retreat to Chattanooga might
be severed entirely. After passing through Hoover's Gap, the 11th Michigan
marched into Manchester on the twenty-seventh, proceeded in the direction of
Tullahoma on the twenty-ninth, and then engaged in a sharp skirmish on July 1
as Thomas attempted unsuccessfully to cut off Bragg's withdrawal short of the
Elk River crossings.[1] Heavy rains and heat plagued the soldiers through day
after day of demanding marches. The 11th went into camp at Decherd Station
on July 8 and relocated to Cowan's Station a month later, patiently waiting for
Rosecrans to consolidate his gains. The Confederates had withdrawn all the
way to Chattanooga, conceding a large swath of southeastern Tennessee almost
bloodlessly. Negley's division was ordered forward in the final movement against
Chattanooga on August 16. Rosecrans aimed to penetrate the mountain gaps
and turn Bragg out of the city, much as he had done with resounding success
at Tullahoma. On September 1 the 11th Michigan crossed the Tennessee River
and resumed its march. Three days later, the regiment halted near Trenton,
Georgia, and surprised and captured several Confederates at an iron foundry.[2]

———

Decherd, Tennessee,
July 12, 1863

Dear Jenny,

We left Murfreesboro the 24th of June and reached this place the 7th of July, coming by way of Hoover's Gap. Thence to Manchester and Tullahoma. We have had no very severe fighting, though the enemy have in several cases stubbornly resisted. At Hoover's Gap they made a stand, but were forced to fall back. I should have written you ere this, but I have had a severe attack of fever. I am feeling better today. The Q.M. was sent to Murfreesboro in charge of a supply train, which left me in charge of the regimental train. Some of the time I was so weak, I could hardly keep my saddle. I cannot write any more at present. James.

Decherd, Tennessee,
July 19, 1863

Dear Jenny,

'Tis Sunday morn, about 8 o'clock in the day. The sun shines very warm at so early an hour. We have moved our camp since I last wrote you, and we are now stopping in a grove of trees, a splendid place for a camp.

I should have given you a sketch of our journeys in my last, but I did not feel like writing. I think I told you, when we left M[urfreesboro], at Hoover's Gap we encamped one day as the Rebels had to be dislodged before we could proceed. There was considerable fighting, but our men did not suffer very severely. Rebel accounts say one of their regiments lost 43 in killed. In the night, the enemy left the hills and retreated, leaving us a clear road. Hoover's Gap is a narrow valley about 3 miles long, surrounded by high hills covered with a dense, heavy growth of timber. The road led through this valley. Had this gap been properly fortified, it would have been a second Thermopylae.[3] About 3/4 of a day's journey brought us to Manchester, a town of considerable size. We reached here about 2 o'clock in the afternoon. Gen. Reynolds's division had entered the town in the morning, and captured about 100 prisoners.[4] We remained here a day or two, and then marched on Tullahoma. The enemy had evacuated the place the night before. There was a long range of rifle pits, probably 8 or 10 miles long. The forest had been felled for miles around to give range to their artillery. We remained in this place one night. The country between Manchester and Tullahoma is the poorest I ever saw, and an unbroken wilderness. I do not remember seeing a single hab-

itation. The first day out from Tullahoma, we made Elk River. The enemy had burned the bridge, and we were obliged to ford. We crossed the train without an accident. This river is not quite as large as the St. Joseph, but [has] a very rapid current. Our road led up the side of a very high hill; when I reached the top I stopped to gaze on the scene that lay spread out as a picture below me. To the north, on the same road we had come, could be seen long white trains. Two or 3 divisions of bluecoats were making their way across the fields. To the east and west, as far as the eye could reach, stretched the green valley, while Elk River in its serpentine course ran through the center. After I had gazed long on this magnificent sight, I turned my horse's head and rode forward at a brisk pace, as the train had gained some distance. We went about 3 miles and encamped, where we remained until we came here. In the meantime we have heard of the defeat of Lee by Meade, the capture of Vicksburg, and fall of Port Hudson, also the rumored evacuation of Charleston.[5] Everything seems to be working for the Union cause. I must thank you, Jenny, for two letters which I have received since we commenced this march. I am regaining my health rapidly. Give my best wishes to your father and mother. James.
Write soon.

Decherd, Tennessee,
July 31, 1863

Busily have I been at work, from early morn until night. I have hardly left my desk only for meals; but Jenny my thoughts have wandered far away from this land of war, and visited the land of peace & happiness. Yes, oft have I thought of you and wondered (Jenny I have not forgotten how to wonder) what you were doing, and then I wondered if other peoples' thoughts ever wandered as far as mine. While my mind was busy in this reverie, I was convinced that such was the case, for the postmaster brought me a letter which plainly showed that someone had thought of me, and that someone was no other than your own dear self. I broke the envelope and speedily read the contents. You described scenes which it has not been my lot to participate in, in a long period. May I someday be permitted to join friends as of old.

You spoke of my folks receiving letters as usual, but you had not received any. There is something strange, Jenny. I have addressed them two letters since coming here, and your own self the same number, at nearly the same date. I told you of my sickness. That I had a severe fever on that long and wearisome march. My health is improving, and I feel nearly as well as ever. I cannot write any more this evening; I am weary. James.

Addison T. Drake.
(Courtesy Archives of Michigan)

I received your letter. You spoke of making Miss Sutton a visit. But you did not tell me what district she was teaching in. Lem is with us again. He came very near dying, but is now as well as ever. Friend Dixon is in good health & spirits. My brother is at Wartrace, Tenn., about 40 miles north of Decherd.[6] Give my respects to your father & mother.

Cowan's Station, Tennessee,
August 12, 1863

Dear Jenny,

We moved from Decherd the 9th, reaching this place the same day about 4 o'clock in the afternoon. The day was very hot, and the men suffered very much. Billy Bournes was sick and came very near giving out.

Cowan is a small town at the foot of the Cumberland Mountains. R[o]usseau's division crossed them yesterday, and we expect to go in a few days. The natural scenery of the place is magnificent. We are camped in a pleasant grove near the bank of a creek that goes dancing o'er its rocky bed, seeming perfectly happy in its winding course. In front on the left and right rise the Cumberland Mountains in all their grandeur. A few days before moving camp, I had paid my brother a visit at Shelbyville. I expected to find him at Wartrace; but on reaching the place found he was at Shelbyville.[7] I laid over one night at Wartrace, and took the morning train for Shelbyville. When I reached there, I found that he had gone out in the country after forage. It was early in the day, and I determined to have a look at the town. I sauntered through the streets until I was tired. The town was busy and lively. Ladies swarmed on the sidewalks in countless numbers. This place has the reputation of being the best Union place in the state. I went to camp, that is where the train was encamped. The boys were getting supper, and when ready, invited me to have their meal. I am too much of a soldier, Jenny, to need a second invite. It was after dark when John came in. He said he should not have known me if some of the boys had not told him I was there. He has changed very much since last I saw him. I had a good visit with him. It was over four years since I had seen him. He came as far as Wartrace the next day with me, and I bade him goodbye and took the 1st train for Decherd.

I have not heard from home in nearly 3 weeks. The boys are all well. Lem is at Decherd with Lt. Drake.[8] Write soon Jenny, and remember me as ever your James.

Cowan's Station, Tennessee,
August 16, 1863

Dear Jenny,

When last I wrote you, I told you we expected to move the next day and cross the mountains; our order was rescinded and we are yet here, though we expect to go in the morning. Major General Negley has just passed our camp. Nearly all the division is on the march today. This is the first marching I have seen done on the Sabbath since Rosecrans came in command of the department. The talk in camp is that we are going to Atlanta, Georgia, though there is not much dependence put on camp rumors.

Now that we have orders to move, the rain begins to fall in torrents. But our road is across the mountains, and very rocky so that the mud will not affect us much.

I have visited Decherd once since leaving. I went to get some money on company savings papers. I saw Lem at Lt. Drake's. He is getting as tough and rugged as ever. I made the trip across the country on horseback.

Write a long letter and tell me all the news. James.

Near Stevenson, Alabama,
August 21, 1863

Dear Jenny,

When last I wrote, we were at Cowan in the state of Tenn.; but now we are enjoying the sultry clime of Alabama. We left Cowan the morning of the 17th and commenced climbing the Cumberland Mountains, which we had to cross in the route. Should I tell you the roads were rough Jenny, I could not tell or convey to you half of the roughness that we had to pass over. I think there was about 7 miles of this very rocky road. These mountains are covered with a heavy growth of forest, of which the walnut, oak, and whitewood are predominant. These first named grow along the base and sides, while farther up along the sides, the cedar is found. After crossing the mountains, our road lay through a narrow valley which wound for miles among the mountains. I do not know when I have enjoyed a trip as well as I did this. The road led through the valley which seemed, or in fact is, walled in from all the rest [of] the world beside. This valley is one vast cornfield. You seldom see any other article raised. Once in a while a small field of tobacco, but seldom. Our camp is in a grove of oak and beech, on the banks of Small Creek, near the Nashville & Chattanooga railroad. Look on the map, Jenny, about one mile north of Stevenson, and you can see where we at present reside. How long we shall remain I cannot say, but shall probably march tomorrow. We are getting rations today. I received your letter yesterday dated the 9th day of Aug., giving me a description of the picnic. I think you must have had a pleasant time. I should like very much to have heard the first orator, as the person spoken of was once a schoolmate. You ask me Jenny to write a long letter, and one about myself. I cannot do this, Jenny. I am not in the mood, and even if I were, I am afraid a soldier is a poor thing to put in a letter. I am the same James I always was, I can see no change. My health was never better. James.

Camp, Near Stevenson, Alabama,
August 23, 1863

Dear Jenny,

This has been a warm and sultry day. I have spent my time in reading. I said reading, but Jenny I have done some thinking. How oft have I thought of home and friends, and wondered how you were passing the day. We are encamped where we were when I last wrote. No orders to move ahead. Preparations are being made to cross the Tennessee. Pontoons have been moving all day, and without a doubt by the last of the week, the main body of the army will be across the river. The railroad is in use between here and Murfreesboro. Trains of cars loaded with provision and supplies for the army is constantly arriving, and the greatest activity prevails. Jenny I am in hopes that in a few short months, this land will be freed from the wicked war which has so long threatened to swallow everything. Already in the distance can I see the light breaking. If our successes for the next three months are as great as they have been in the past, there will be no Confederacy. Write soon and believe me as ever your James. Friend Dixon is in the tent, Lem is well.

Camp,
4 miles from Trenton, Georgia,
September 5, 1863

Dearest Jenny,

I received your most welcome letter of August 23d, but have delayed answering on account of being on the move. We left our camp at or near Stevenson, Ala., the 1st day of September, about four o'clock in the afternoon. Passing through Stevenson, we reached the Tennessee River at 9 o'clock. It has been very dark in the early part of the evening, and we had to await the rising of the moon before it was considered safe to cross the pontoons. We succeeded in crossing without accident.[9] After this, we went four miles and camped for the night. The Tennessee is the most beautiful I ever saw. Its banks are high and covered with a forest of heavy foliage. On our march there has been nothing exciting. No Rebels have appeared to stop our progress. The country is very mountainous, and but few inhabitants. I have told you so often of rough roads, Jenny, that there is no use describing this. It was rough, very rough.

Our destination is not known, but it is evident that we are in the rear of Chattanooga, and if the Rebels stay at that place, we shall go north before going south. If they have fallen back from there, we will go to Rome or Atlanta.

Write soon Jenny, and remember me as ever your loving James.
Friend Dixon is sick with fever, though not dangerous.

Negley's advance resumed toward LaFayette, Georgia. On September 8 the 11th Michigan became the first Federal regiment to enter McLemore's Cove. Negley's lone division brushed aside Confederate cavalry and pressed ahead through the valley separating the looming Lookout and Pigeon Mountains. Braxton Bragg, who had elected to abandon Chattanooga, recognized the blue division's isolation and dispatched orders for his subordinates to swallow Negley's force whole. But Bragg's subordinates approached timidly, and Negley realized his predicament before it was too late. Near Davis's Crossroads on September 11 the Federals, bolstered by two freshly arrived brigades, executed a fighting withdrawal during which Stanley's brigade of a thousand bluecoats was briefly called upon to fight a delaying action against seventeen thousand advancing Confederates. The 11th Michigan fought with determination, losing three dead, thirteen wounded, and three missing.[10]

The opposing armies joined again in battle near Chickamauga Creek eight days later. Negley promptly dispatched his division train, with the exception of the ammunition and ambulance wagons, to Chattanooga. Quartermaster Sergeant

Chickamauga by Alfred R. Waud. (Library of Congress)

James King felt conflicted about leaving his friends in a time of danger, but he left with absolute confidence that his comrades were prepared for the coming shock:

> The men composing this brigade had been familiar with the rifle from boyhood, and they were veterans of two years of active service in the field. They had been thoroughly tried in the crucible of battle. The 11th Michigan knew it could rely upon the 19th Illinois and the 18th Ohio to do their full share of the terrible and difficult work before them, and each soldier had perfect confidence in his file companion. . . . It would have been hard to have found a brigade in the Army of the Cumberland, or in any other army, that ever marched into battle with more confidence than did this little brigade on this historic field of Chickamauga.[11]

Stanley's brigade saw little action on September 19, but on the twentieth the unit fought with distinction, delivering two powerful blows that helped blunt key Confederate assaults. Late in the morning, as the Rebels flanked the Union left, threatening the destruction of Rosecrans's army, Stanley's brigade arrived near

Kelly Field just in time to ambush the charging Confederate brigade of Brigadier General Daniel W. Adams, scattering the Rebels with severe loss. The 11th Michigan took General Adams prisoner. That afternoon, Stanley was wounded and Stoughton rose to brigade command. Colonel Stoughton led the unit through its finest moment, the last-ditch defense of Horseshoe Ridge, or Snodgrass Hill. A breech mistakenly opened in the Union line to the south had resulted in one-third of the Union army fleeing the field, and it fell upon George H. Thomas to avert the complete loss of Rosecrans's host. The 11th Michigan, with just two hundred men left in its ranks after taking casualties and detaching prisoner guards for Chattanooga, confronted ferocious assaults by more than five hundred Confederates from Archibald Gracie's brigade. The tenacious Wolverines cut down close to two hundred Rebels in desperate fighting that went hand to hand at times. After dark the Union army, battered but intact, retreated to Chattanooga.[12]

Stoughton's brigade arrived at Rossville, on the road to Chattanooga, around midnight. On the way, the Wolverines were happily reunited with James King. "Here," Lieutenant Borden Mills Hicks of Company E remarked, "our Quarter-Master Sergeant met us with needed supplies, and what was better than even something to eat, letters from the loved ones at home. . . . After preparing our midnight meal, and drinking freely of the clear cold water from the creek, we layed down on our arms to get a much needed rest." The 11th Michigan counted five of its soldiers killed. Forty-two were wounded and nineteen missing. On the next day, Stoughton's brigade was among the units comprising the Army of the Cumberland's rear guard at Rossville Gap. "No greater responsibility than this ever rested upon a brigade commander," James King asserted, "for a failure in the performance of that duty would have involved the destruction of the Army of the Cumberland. In conversation with [Colonel] Stoughton in regard to this achievement, he said: 'I received my final orders from Gen. George H. Thomas, and I know he regarded it as a dangerous and important duty, and he complimented me personally on my success.'"[13]

Chattanooga, Tennessee,
September 23, 1863

Dear Jenny,

When last I wrote you, we were south of this place in the state of Georgia. The enemy received heavy reinforcements, and our army has fallen back and are now busily engaged in fortifying. The fighting has been very severe; that of Sunday the most desperate. Our loss is heavy, that of the enemy greater, as our men fought behind breastworks which they had constructed of stone and rails.

Charles W. Newberry. (Belknap, *History of the Michigan Organizations at Chickamauga, Chattanooga, and Missionary Ridge*)

Several of our regiment were killed, but none of the boys of your acquaintance. The captain of Co. C, Charles Newberry, was killed instantly.[14]

Lem has gone to Bridgeport for supplies. Dixon is well. Billy Bournes was left at Cowan sick. My brother was here, but went to Bridgeport in charge of a train.

We expect to be reinforced by a corps from Grant's army.[15] Everything [is] quiet this morning. My best wishes to your father & mother. James.

Chattanooga, Tennessee,
October 4, 1863

Dearest Jenny,

'Tis not because I had forgotten that I had friends who were anxiously awaiting letters from me that I did not write, but the reason is this. The mails are very irregular and one can hardly know when a letter will reach friends.

Jenny, a great battle has been fought, and our army has gained a victory, though I see by Northern papers that it is considered a defeat. In the first place, our general aimed at the taking of Chattanooga, causing them to evacuate east Tenn. and virtually clean Tennessee of Rebels. This was all the army expected, and as high as public expectation ran. This has all been accomplished, and at very little sacrifice for the amount of work done.[16] In the first place, two corps of Rosecrans's army was sent into Alabama to the right of the place, making a flank move, causing the Rebels to evacuate their stronghold and fall back. Crittenden's corps there occupied the town without resistance. I think I wrote you when in the valley some 26 miles south of this place. If you remember, our division was in the advance. We lay in this valley a few days when the whole corps, which is commanded by Maj. Gen. Thomas, came up with us.[17] I think I wrote you of the fighting of our division as soon as we entered the valley.[18] Here the corps of McCook, who was to our right, joined us, and we commenced to cross the valley in a northeasterly direction. We went probably 8 miles when we came up to Crittenden's corps and joined forces with him. This comprised the Army of the Cumberland. Our line of battle ran in a northeast and southwesterly direction, the enemy being in our front and in the direction of [the] Pigeon Mountains. We lay here all day and night until the next afternoon. (You must remember Jenny that Bragg was heavily reinforced after the evacuation of this place.)[19] It was about 3 o'clock, I think, when the enemy made a furious attack on our left, that is Crittenden's corps, and our division and that of R[o]usseau were sent to support them. We moved away to the left, but our division was not engaged. The next morning we received orders to move our train to the city. This was Saturday. We could feel the throb of cannon and hear the roar of musketry as we left. We reached the city that day, and on the morrow I was sent to the front with a team loaded with rations for the men. We proceeded on the road about 3 miles when the report came that our army was all cut to pieces. This news we got from stragglers. We were ordered to stop. We lay here from noon until dark while down the valley came the boom of cannon, and roar of musketry. It was Stone River enacted again. Just as darkness set in, we received orders to move forward. I reached the regiment about 9 or Ten o'clock. They had fought all day against appalling odds and had held their ground, in fact the whole army had done the same. But none so hard as the center corps commanded by the gallant Thomas. They were in the best of spirits, and had lost lightly considering the desperate fighting which had taken place. The Rebels acknowledge a defeat this day, as they had nearly two to our one killed. Our men had fought behind breastworks which they had constructed during the night before the battle. I left the front Monday forenoon with the team. There was no severe fighting this day, only in the afternoon, when the enemy made several attacks but were repulsed in every instance. That night our army withdrew and fell back into their stronghold from which the generalship of our Rosecrans had drawn them.

Now Jenny, this cannot be a defeat when they lost nearly double our number and were unable to follow up their great victory which Northern papers are whining over. If we had made a direct advance and had 30,000 men slaughtered and merely occupied the place, it would have been called a great victory. The amount is that we have gained a great work while they have lost, and confess can never be repaired without the occupation of this place. This they never can do. They never can drive us from here.[20]

The worst blow which we have received is the burning of a train of supplies between here and Bridgeport.[21] Our regimental teams were among them. Billy Davis was in charge of our train, but made his escape. I should have went with them, but it being the end of the month, the quartermaster said I had better remain and make out the reports. Many of the teamsters were taken prisoners, and some of them killed. Billy was back today at the spot where they were captured, and everything is a perfect mass of ruins.

I received your letter today, and am truly thankful. It is some time since I have heard from you. The boys are all well. Lem has gone to Bridgeport for supplies. I cannot write more at present. From your faithful James.

The enemy kept up a continual cannonade all day but done us no injury. [Washington] Irving Snyder of Three Rivers died last night of a wound received in Sunday's battle.[22]

Oct 5. Our men have opened fire on the Rebels from the fort, but they do not reply.[23] Their tents and wagon train are plain to be seen from our camp.

––––––––––

Despite their valor at Chickamauga, Stoughton's men were fated to share in the Army of the Cumberland's humiliation, starving under siege in the dearly won city of Chattanooga. The Confederates put a stranglehold on the roads leading into town. By October 8 the delivery of provisions and materiel required the traversal of a grueling sixty-mile route winding through the mountains.[24] With food scarce, horses and mules perished by the thousands. But Bragg's siege was soon cracking at the seams. Ulysses S. Grant assumed command of the newly created Military Division of the Mississippi, which combined Burnside's Department of the Ohio, Rosecrans's Department of the Cumberland, and Grant's Department of the Tennessee, under a single command. Grant promptly replaced Rosecrans with George H. Thomas and arrived in Chattanooga in person on the twenty-third. He soon ascended Cameron Hill, accompanied by Thomas. The 11th Michigan and 19th Illinois were encamped on the very same prominence. Standing within earshot, James King recorded the scene.

> Gen. Thomas spent nearly an hour in pointing out the exact situation of the two armies. The writer of this remembers the great commander, Grant, at that time as a man of medium size, light complexion, full beard closely trimmed and clean-cut features as were ever placed on

the face of mortals. He was dressed in citizen's clothes of a brownish hue, a slouch hat, cavalry boots well bespattered with Tennessee mud, and nothing on by which you could designate his rank. He had ridden eight miles on horseback that morning from Brown's Ferry over roads not much better than a mortar bed, and that accounted somewhat for the dilapidated appearance which he presented alongside of that beau-ideal soldier, Geo. H. Thomas.

The presence of these two noted generals attracted the attention of all the men in camp, and they gathered around as closely as they dared in order to see and hear what was going on. It may have been the looks of the veterans of the 19th and the 11th, or that in part, which [later] caused Gen. Grant to say to Sherman that "Thomas' army were so starved that the men in hunger stole the few grains of corn that were given to the favorite horses; that the men of that army had been so demoralized by the battle of Chickamauga that he feared they could not be gotten out of the trenches to take the offensive, and he wanted Sherman's troops to hurry up and take the offensive first, after which he had no doubt the Army of the Cumberland would fight well." Right there before him, however, were the veterans who had stood with Thomas on Snodgrass Hill in the fiercest tempest of bullets ever encountered by mortal man since the use of firearms began.[25]

Grant immediately set about prying Bragg's fingers off the supply line. Joseph Hooker had arrived in the theater earlier in the month with the 11th and 12th Corps from the Army of the Potomac, and Grant wasted no time putting them to work. Hooker and Thomas successfully cooperated to open a lifeline out of the city. Less than one week after Grant's arrival, materiel and rations were flowing into Chattanooga.

———

Chattanooga, Tennessee,
October 13, 1863

Dear Jenny,

Another week has passed away, and we remain in quiet possession of the city. The place which, of all others, the Confederacy desired to hold. The land which His Majesty Jefferson [Davis] said nearly a month ago must be wrested from the hands of the invaders, as the glorious victory which they won in Georgia would amount to nothing. They have not wrested from us the coveted treasure, and his saying has wisely come to pass. We have gained all that we worked for, and they have lost. I am in hopes that it will not be many days before we will be able to work on the offensive again. The enemy confronts us, his pickets are in plain view from the fort. There is an agreement that no firing between pickets shall be allowed, and day after day they stand watching each other in close proximity to our camp.[26] Once in a while a party will venture in view from the Rebel camps, but a few words from our 12-pound gun on the fort tells them to keep a respectful distance. 'Tis a rainy, dreary day Jenny, I have been a close occupant of my tent all day. The rain is falling in copious showers, and dark clouds obscure the bright sun. 'Tis the right kind of a day to be melancholy if I was a melancholy individual, but as I am not, I listen to its ceaseless patter and think there is a day of brightness drawing near. Why should I be sorrowful Jenny, when I have friends who are anxiously awaiting my return, and every passing hour lessens my stay? 25 months of my enlistment has passed away, and I am certain ere the next eleven expire, peace will reign once more through our now distracted country. Then, Jenny, it will be sweet to return to friends and home. I cannot write more. Write soon. James.

October 15. Two days have passed since writing this, and I have had no opportunity of sending it to Bridgeport, as the mails are very irregular. There is nothing new, only the report that the enemy are receiving heavy reinforcements.[27] It has been rainy for the two past days, and still continues to pour down. The river is rising very fast. Lem returned from Bridgeport today. He is nearly worn out with exposure, not having had any rest since we occupied the place.[28] Dixon is well, and as full of fun as ever. Jenny I have not had any letters in some time.

Chattanooga, Tennessee,
October 30, 1863

Now Jenny I have nothing to do today but write you a letter; though I have nothing new to tell. Even if I should tell you that the 11th Army Corps under Hooker had a heavy fight with Longstreet near this place and captured a number of prisoners, it could be no news as you would see it in a half dozen papers ere this reaches you; and that the Rebels had been using artillery from Lookout Mountain but had done no damage, this would not be news.[29] Then I will write something which is not in the papers. 'Tis raining Jenny, not one of those gentle pattering showers which is spoken of in the song, but a shower such as they had in the days of Noah, a regular deluge, though I do not think it will continue forty days and nights. No, in this clime it rains a day or two, just enough to get Mother Earth terribly mixed up, then the bright sun appears

Chattanooga and Lookout Mountain. (Guernsey and Alden, *Harper's Pictorial History of the Great Rebellion*)

and dries the mud so that the employees of Uncle Samuel can begin to think of active work, when another storm sets in and we are forced to stay inside our tents or brave the dangers of being mixed in the deep and heavy mud.

I received your letter written the 11th of October, the 3d letter I have received since leaving Stevenson, a period of two months. Now Jenny what do you think of this? One who is so far from home and flatters himself that he has so many friends, and receives only 3 letters. I must give you credit of writing two. In your letter you spoke of Isaac Stell being wounded in the Battle of Port Hudson. I had forgotten that he enlisted in the 6th Michigan until you mentioned his name. What has become of his brother Benjamin who enlisted in the 2nd Regt.? I think that the regiment belonged to the 9th Army Corps and are with Gen. Burnside in east Tennessee.[30]

You spoke decidedly in favor of the draft. If there was a draft of six hundred thousand men made immediately, it would be the quickest means of ending the war. We shall probably see our term of enlistment expire in the service. It does not seem so long a time since leaving home and friends, but over 26 months has passed away, and not quite 10 to serve.

Friend Holbrook has been discharged on account of poor health, and will return to Michigan in a few days. He has proven a true friend to me indeed.[31] He says if he returns to Michigan he will visit my father's family.

I cannot write more at present. Give my best wishes to your father & mother. Write often. James.

Chattanooga, Tennessee,
November 20, 1863

Dear Jenny,

I received a letter from you a few days since, the first I had in some time. I shall not attempt an answer tonight, as I have not time to write a long letter.

There is nothing new, but in a few days the army must again fight or fall back.

Excuse haste Jenny, when I have time, I will write lengthier letters. Keep up good spirits, and may God bless you is the earnest prayer of him who thinks often of thee. James.

Chattanooga, Tennessee,
November 23, 1863

Dear Jenny,

There has been some heavy fighting today. The left of our line has been advanced considerably, and in their advance captured a number of prisoners.[32] Our regt. are on picket at the foot of Lookout Mountain. None of our men were injured, except a Company B Boy who was hit by a piece of shell. I came to camp a little while ago and everything was quiet there. We are confident of success. The enemy's batteries on Lookout Mountain and Missionary Ridge were handsomely worked. Our artillery firing, splendid. Moccasin Point battery [has] complete range of the enemy's camps and Lookout, while the guns from Forts Wood and Negley caused much commotion among them to the east. I am weary, and will wish you a kind goodnight. James.

––––––––––

On November 24 Hooker's bluecoats famously charged up mist-enshrouded Lookout Mountain, the anchor of the Confederate left. That night, James King correctly anticipated the following day's frontal assault against Missionary Ridge. He secured permission from brigade commander Colonel Stoughton to grab a rifle and join the ranks—going over the head of regimental commander Major Benjamin G. Bennett, who, upon hearing the news, predicted his own death in the charge and begged unsuccessfully for James to stay behind with the wagon train where he belonged.[33]

On November 25 fresh reinforcements under William Tecumseh Sherman were brutally repulsed in their attempt to flank Bragg's right, and Grant sought to relieve the pressure on Sherman's front by ordering Thomas to seize the first line of rifle pits at the base of the ridge. By 4:00 P.M., the Army of the

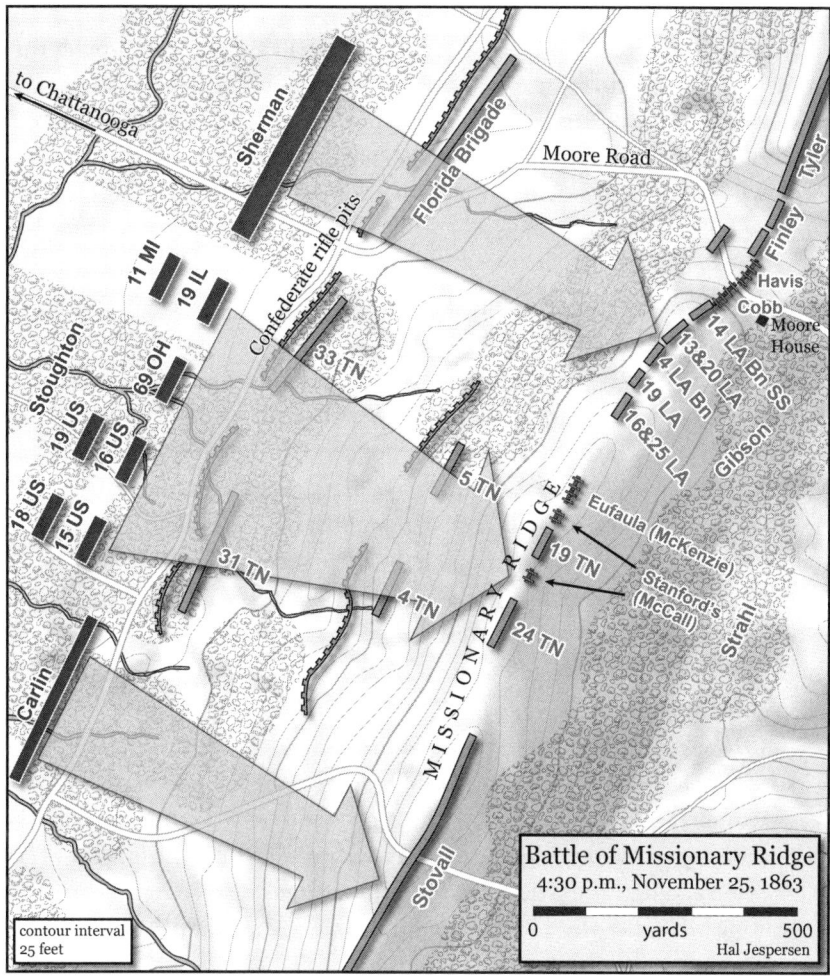

Battle of Missionary Ridge
4:30 p.m., November 25, 1863

0   yards   500

Hal Jespersen

contour interval
25 feet

Cumberland was charging across the expansive half-mile of open field before the ridge, under heavy artillery fire. As Stoughton's troops moved to engage the Confederate brigade of Otho F. Strahl at the base of the heights, James King—who should have been miles away from the shooting—was observed by several of his comrades to charge ahead of the brigade. He became one of the first bluecoats to enter the Rebel works. Bayonets were employed in the brief melee that ensued, and numerous butternuts were captured. After a brief pause, Thomas's troops realized they were vulnerable to fire from the ridge above and resumed their charge uphill against orders, facing a vertical climb of about five hundred feet over the distance of more than half a mile—a taxing climb under the best of circumstances, let alone in the face of an entrenched, veteran enemy. Once again, the men of the 11th Michigan watched as James King rushed ahead of Stoughton's unit, becoming "one of the leaders through this storm of death and destruction," as Borden Hicks put it. James was among the very first Union soldiers to crest the ridge. Just as Stoughton's advance faltered below, James charged into a small gap mistakenly opened in the Confederate line on the summit, with a trickle of other bluecoats following in his wake.[34] James recorded the battle's climax:

> In a moment after we had gained the top of the ridge [we] could see the enemy seeking safety in flight down the eastern slope. Just to our right the Rebel line was still intact, and not more than five or six rods away was a semi-circular earthwork where [there] was a battery of six guns, which were being worked on the Union troops farther down the ridge. There was also a support of four or five hundred Confederate infantry who kept up a murderous fire on the troops below. A chestnut tree, about twenty inches in diameter, stood on the ridge where it began to slope to the east and lying beside it was a log nearly as large. The two formed a niche, and I was not long in taking advantage of the safety it afforded. Lying in this nook I would load lying on my back, then turn over and fire across the log at the Confederates who were working the guns. Every time I showed myself above the log Minie balls would whistle over me or strike the log at my right. I had loaded my gun lying on my back for the twenty-first time, and on rolling over to cock and cap it, I raised my right arm a little too high, a Minie ball struck it, breaking the bone just above the elbow joint. I could do no more fighting on my own hook, and after waiting behind the log for about a minute, the enemy thinking perhaps they had killed me, I jumped and ran back about a rod to a large tree, behind which was Byron Liddle, of Company D. He had just fired his gun and I told him to cap mine and give them another shot. About this time we had made it so uncomfortable for the Confederates from our flanking fire that the infantry and artillery men fled precipitately down the eastern slope to the ridge, leaving the guns in the works. When I had time to look around for the rest of the comrades I found a little to my left Captain B. M. Hicks and about one hundred of the 11th with the regimental colors, and a few rods away was Sergeant B. F. Hart with a score or more of the regiment, who had gained a lodging just outside the Rebel works, and they had also been keeping up a telling fire on the Rebel gunners. I had my comrade Liddle tie a handkerchief around my arm to prevent loss of blood and then I hastened to Captain Keegan, who

Borden Mills Hicks. (Courtesy Clements Library, University of Michigan)

Benjamin Grove Bennett. (Belknap, *History of the Michigan Organizations at Chickamauga, Chattanooga, and Missionary Ridge*)

The Battle of Missionary Ridge. (Guernsey and Alden, *Harper's Pictorial History of the Great Rebellion*)

had taken command of the regiment when Major Bennett was killed, and told him to hurry the rest of the regiment forward, that a lot of the boys had broken the Rebel line on the top of the ridge, that the enemy had deserted their guns, but that they might return and that they would need help to hold them. When I reached the top of the ridge for the second time the place was swarming with troops and there was no danger that the enemy could repossess their former ground. Besides the troops of our own brigade at this time I remember the Forty-Fourth Illinois, for Colonel Barrett was a personal friend of mine, and we shook hands within a few feet of where the Rebel guns were standing. He seemed almost crazed with joy at the unexpected turn things had taken, and at the grand and heroic work which the Army of the Cumberland had just performed. He said: "Look back over the ground, think of the deadly fire that was poured upon us; it is a miracle, and this will be looked upon as the grandest military charge ever made since the world began."[35]

James would never realize it, but the exploitation of this gap in the Rebel line had triggered the rout of Alexander P. Stewart's entire division. After watching the Confederates disappear from view, James marched three miles back to camp, dropped off his equipment, and trudged another mile to the general hospital in Chattanooga, where the surgeons immediately prepared to amputate his arm. At that moment James had only one thought: "What will Jenny think of a cripple?" But regimental surgeon William Elliott intervened and saved the mangled limb, earning the wounded soldier's eternal gratitude.[36]

Long days and sleepless nights of suffering ensued while James convalesced, but he experienced a joyful event that mitigated his anguish the day after the battle: a visit from his brother. John King had taken a bullet to the forehead charging up Lookout Mountain, but the projectile was of small caliber and drilled clear through the brim of his slouch hat before flattening on impact with his skull. His scalp bled freely, but the Illinoisan suffered no permanent injury. John remained with his regiment and beheld the assault on Missionary Ridge the following day from the very summit of Lookout Mountain. "Fiction," James commented, "could never picture grander or more stirring events than those through which these two sons of Michigan passed; neither could it portray a more joyful meeting than took place between them in that hospital, even among hundreds of wounded and dying." On January 23 James, still unfit for duty nearly two months after being shot, was granted a thirty-day medical furlough. He returned to Fabius, Michigan, where he was reunited with friends, family, and Jenny, after an absence of more than two years. On February 15 he requested an extension to his leave of

William N. Elliott. (Courtesy Alyssa Chandler)

absence, informing Colonel Stoughton that "my arm is very weak, but [I] think [it] is improving." Yet the quartermaster sergeant departed for Georgia at the end of the month. Some in Fabius believed that James couldn't wait to return to the war. His example seems to have been contagious: at about the same time he left Michigan, Jenny's father volunteered to join the 6th Michigan Heavy Artillery.[37]

Rossville, Georgia,
March 6, 1864

One week ago today Jenny, I was enjoying the society of friends near and dear to me in my own loved home; today finds me among the scenes of camp, far from home in the sunny South. I left White Pigeon Monday evening at 5 o'clock, and the next evening before five I was in Louisville, Ky. I stayed all night in that city, and the next morning took the Nashville train. After we had went about 50 miles, the train came to a halt on account of the train in advance of us meeting with an accident. The engine was capsized & cars were off the track, and things were pretty generally mixed up. This kept us waiting until 12 o'clock that night. After leaving Stevenson, another misfortune occurred to the train in advance of us. The whole train was thrown from the track, the engineer and fireman killed, so you see that railroading is not very safe in the South. I finally reached Chattanooga Friday noon. I stayed at that place until this morning, when Webb [Daniel Webster Holbrook] came to the regiment with me.[38]

Confederate prisoners at the Chattanooga railroad depot. (Library of Congress)

The weather is very pleasant, like summer. I send you some peach blossoms. I was afraid if I should tell you this without proof that you could hardly credit it.

Things in the regiment have not changed much since I left. The boys are all well. Billy Davis, [the] quartermaster, and Harkness are in the tent with me. Lem is improving in health. He is staying with Lt. Drake.[39]

Jenny, you may think that I wished to get away from home and had wearied of the scenes which there surrounded me. But 'twas not so. The first time I left home was not near the trial which it was this time. I know that some thought I valued the scenes of home lightly, but 'tis not so. I cannot write a long letter, am not in the mood. Write as soon as you receive this dear Jenny, and tell me all the news. Have you heard from your father yet? Remember me kindly to your mother & Darius. James.

Rossville, Georgia,
March 9, 1864

My Dear Jenny,

Now I have seated myself at my desk to write you a letter, what shall I write? Shall I tell you that I visited the city of Chattanooga today, made the colonel and Holbrook a visit, and had a good time generally? Stayed to dinner and returned just in time to take supper with the boys. We did not have hardtack for supper as you would suppose, but our very intelligent American of African descent made us some warm biscuit which would have been called eatable in a land where people pretend to live.[40] The other articles I will not name, but consider them too numerous to enumerate.

The day has been one of sunshine, but tonight the rain is falling and comes in fitful gusts against our house of cloth. Well, let it rain Jenny, I shall not complain. Could you see the inside of our tent, you would pronounce it comfortable. The tent is a large one. A Sibley stove stands near the entrance. My desk is against one side of the wall, about midway of the tent. The quartermaster and Captain [Ephraim] Hall are playing at cards near the desk. This last feature you may not like, but Jenny, although I do not play the game, I can see no harm where they play for amusement, any more than in any other game that is played.

The contrabands in their tent are having a fine time, to judge by the way they laugh. Jenny there is four times the mirth in their dark skin than in the white man. They seem to have no cares.

I hope, Jenny, you will write often and tell me all the news.

Tell your mother that I should like to hear from her, thus a letter would be very acceptable. With many good wishes, I am as ever your faithful James.

Chattanooga, Tennessee,
March 18, 1864

My Dear Jenny,

Mine has been the privilege today of standing on the point of Lookout Mountain, of which you have heard me speak so often. As I stood and thought of the long and fearful struggle which gave the very rocks on which I stood into our possession, and thought of the many weary months of suffering which we were compelled to undergo while it remained in the possession of the enemy, my thoughts were those of exultation. Although it was purchased at the fearful cost of blood. Let me, Jenny, try and give you a general idea of the feature of this ragged and massy pile of rocks which lifts its head far in the air, overlooking all the surroundings.

This is the terminus of a long range of mountains which extends far into Georgia. The top of the mountain is slightly undulating, and covered with trees of many varieties. Among these is the celebrated laurel. The sides are rugged and craggy, rising from seventy-five to hundreds of feet perpendicular in height, and then gradually losing itself in a gentle slope to the east almost two miles away in the valley of Chattanooga, and to the west in the valley of Lookout. The point upon which I stood is a single narrow rock eighteen hundred feet above the river, which winds in a serpentine course at its foot, and just two miles away from the fortifications at Chattanooga. Far away in a northeasterly direction, you can trace the windings of the Tennessee for miles and miles as it winds around among the mountains of east Tennessee. Range upon range of mountains, peak after peak, meets the eyes until you get tired of gazing. The peaks of North Carolina, Virginia, and Kentucky can be seen, and farther to south is a peak which is said to belong to the state that passed the first accursed acts of secession. Now Jenny, we have been viewing the country at a distance, let us cast our eyes on scenes which I am acquainted with. We will partly turn around, facing the southeast. Far away, perhaps 30 miles, can be seen a range of mountains touching the blue sky. These are the Pigeon Mountains of which I told you in one of my letters written before the Battle of Chickamauga. You remember my telling you of General Negley's attempting to cross this range of mountains, but meeting a superior force of the enemy, was compelled to fall back and await reinforcements. It was behind this range that the troops of the Confederacy from every state and army massed to crush the Army of the Cumberland. There are several gaps in this range, and out of these on the morning of the 19th of September they swarmed like a pack of hungry wolves. The valley of Chickamauga is spread out like a map before us. Here Jenny one can say truly that destruction has swept the lovely plain. The houses are nearly all destroyed, the fences swept away, the country depopulated. These are the sad realities of war. But Jenny I am getting tired of attempting a description. Oh how I wish you could gaze on the scene. I am sure you would enjoy it.

The regiment moved to Graysville a few days since, a place about 14 miles southeast of here.[41] I have been stopping with Lt. Drake a few days, and shall go to camp tomorrow.

The roads in general are in good condition. Give my best wishes to your mother and the rest. Has your father left the state yet? Yours faithfully, James.

Jenny's cousin, Augustus Milo Wellman. (Courtesy Sue Boyer)

Graysville, Georgia,
April 7, 1864

My Dear Jenny,

Your welcome letter written the 24th of March came to hand yesterday, and I hasten to reply. I had looked some time for the welcome missive, and at last it arrived. You speak of my visit at home and the fun, happy hours and days I spent with loved friends. Yes indeed, they were joyous moments, never to be forgotten by me as long as reason holds its sway. Who can appreciate the comforts of home and the society of friends better than one who has been absent two and a half years and had to undergo the temptations, exposure, and dangers of a camp life? When I shall again return I cannot tell; but Jenny wherever I go or whatever may be my surroundings, I never shall forget the one who has loved me three and a half long years. Many are the things that I have seen in life. I have heard you speak of these changes many times, but there is no danger in our case. You did doubt me once, but Jenny I do not think you were to blame, considering the circumstances.

You speak of having plenty of snow since my leaving. What if I should say that we have had enough in this sunny land to make good sleighing? Mother Earth was covered with a mantle of white several days. This was rather chilly after seeing the buds and blossoms of spring. The bright sun is with us once again, and our visitor in white entirely disappeared.

It is provoking that our habitation is so far away that we cannot entertain visitors, but I am afraid if you should make us a visit, you would not think soldiers very good housekeepers. You have, I suppose, heard the song of "Bachelor's Hall what a quare-looking place it is."[42] I think a soldier's housekeeping corresponds very well with this description. Camp kettles and pans, and other things too numerous to mention, make up the articles and furniture of our homes of canvas, and you at home are not used to seeing one room made into a kitchen, parlor, sitting room, and sleeping apartment. Our seats are benches and boxes. Wait until your father has learned the uses and ways of a soldier's life. I think he could describe them better than I. Jenny I often think my descriptive powers poor. You did not tell me where your father had been since I left, and where he has to report at the expiration of his furlough. I understand that the 6th Mich. is at Coldwater, Mich. How long have they been in the state, and when do they return south?[43]

I have not heard from sister since my return, and since the receipt of your letter I have felt very anxious on her account.

[Your cousin Augustus] Milo Wellman reached the regiment today and has been assigned to Company A. He left Grand Rapids last week.[44] There is no truth in the report of the regt. going to Texas. Your letter was the first I had heard anything about it. I received a letter from friend Holbrook today. His health is very poor. He said Lemuel was feeling very well, for him.

Remember me kindly to all. James.

On April 11 James was the subject of a Certificate of Disability for Discharge, filed by Adjutant Linus Truman Squire. The form was approved and forwarded by Captain Ephraim Hall, who was temporarily in command of the 11th. The certificate continued on up the chain of required approvals, receiving nods from medical personnel at the division and corps levels. Next, the Medical Director's Office of the Department of the Cumberland proclaimed James disapproved for the Veteran Reserve Corps. The Reserve Corps provided duties for convalescent soldiers, far away from the front lines. Mysteriously, the next and final endorsement on the form, penned by Henry Cist of George Thomas's staff on April 22, persisted in dictating that James "be transferred to the Veteran Reserve Corps. . . . By Command Maj. Gen. Thomas." Yet James stayed the course with his regiment, and made no mention of this in his letters to Jenny. Did the sergeant ask to stay with his comrades even though his health dictated otherwise? The same discharge certificate contained a medical statement from Dr. Elliott, diagnosing limited mobility of the elbow and declaring James partially disabled. The young man's arm was permanently shortened by the loss of bone. There was muscle atrophy, the limb could not be straightened, blood circulation to the forearm was poor, and varying levels of discomfort would linger as long as he lived. His arm would never again endure the kind of manual labor that had been a way of life on the farm.[45]

CHAPTER 6

The Cannons' Deep Roar

April 1864–December 1865

———— ∞∞ ————

I think I have done my duty,
and further than that I will not go.

When the time came to consider reenlistment, James was influenced by at least two, and perhaps three, reasons not to renew his commitment to the army. First and foremost, Jenny was pressuring him to return home. Second, his arm was not up for the rigors of continued service. A third likely factor—though he did not comment upon it—is the fact that he was surrounded by friends and respected peers who had no intention of staying in the army. Strikingly, the desire to reenlist was all but nonexistent in the 11th Michigan Infantry. More than half of all Union soldiers eligible for the veteran reenlistment bounty and furlough took advantage of the offer, yet in the 11th Michigan, a mere fifteen soldiers followed suit. By way of comparison, more than one-third of Michigan's infantry regiments achieved the 75 percent reenlistment rate required to retain a unit's identity. Of all the individuals in the 11th Michigan mentioned by name in this book, only Byron Liddle opted for veteran reenlistment.[1]

Ulysses S. Grant had been summoned out east, promoted to lieutenant general, and appointed to overall command of the United States armies. He attached himself to the Army of the Potomac, and William Tecumseh Sherman was assigned Grant's former command in the West, placing Thomas and the Army of the Cumberland under his orders. Grant and Sherman would synchronize their 1864 campaigns, denying the primary Confederate armies any opportunity to reinforce each other.

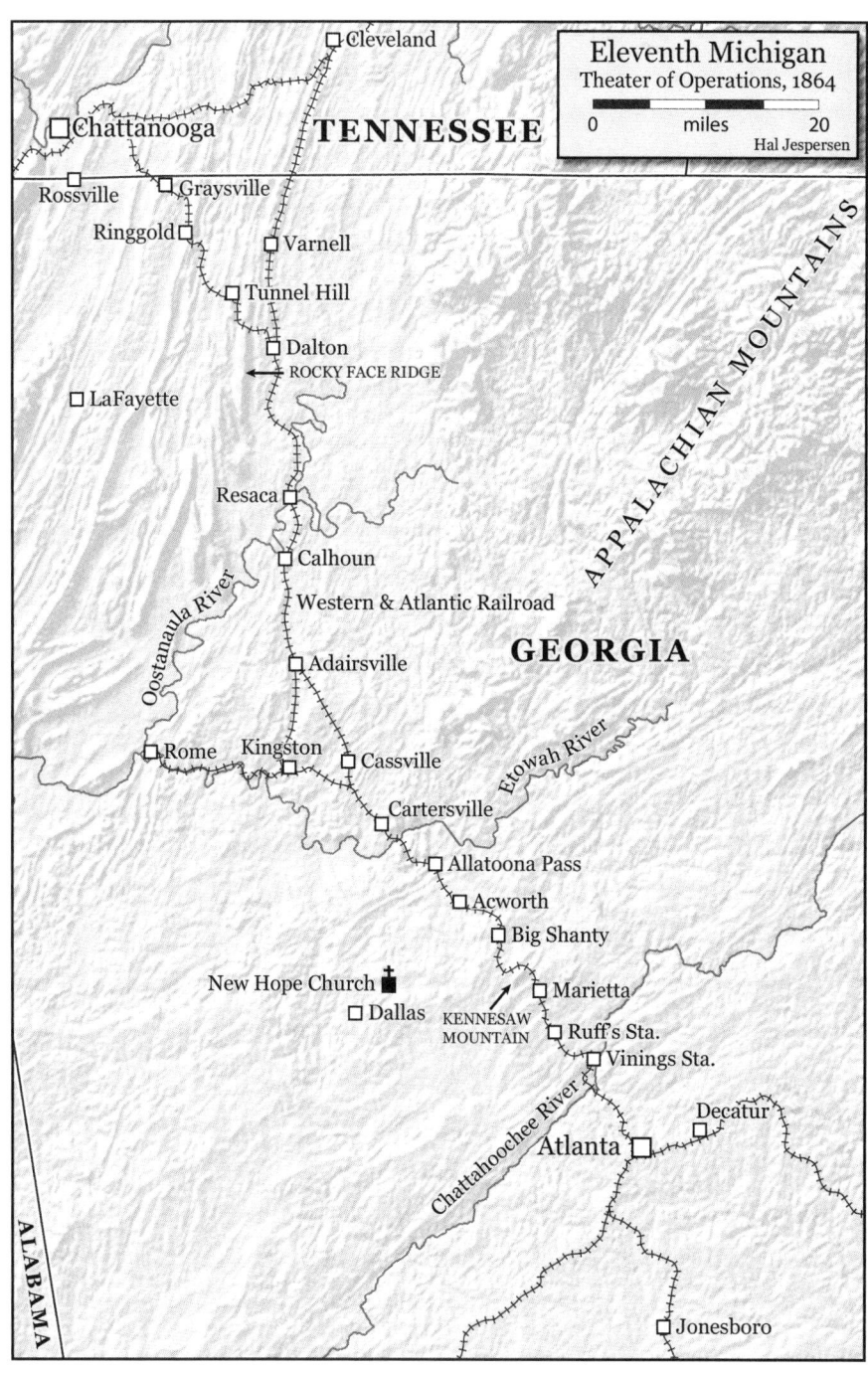

Eleventh Michigan
Theater of Operations, 1864

0 miles 20

Hal Jespersen

Graysville, Georgia,
April 12, 1864

My dear Jenny,

This has been a rainy, drizzling, and disagreeable day. At early dawn the sky was clear, but before 9 o'clock the bright sun was overshadowed and the rain quietly falling, but what care I for the rain. I had enough to keep me busy. I have been issuing clothing to the companies. Once a month this issuing has to be gone through with. The companies make their estimates for the month to the regimental Q.M. He makes a requisition on the brigade Q.M. for the articles required, and thus you see the way the companies are supplied.

I am in hopes that the fair weather will make more activity in our armies. I do not think that we shall make a movement far south from this quarter. True, there are troops coming in, but not an army large enough to fight on the offensive.[2] If a movement is made on the Potomac we shall probably hold our line here, and may advance as far south as Dalton. I cannot but hope Jenny that the war will be ended the coming summer.

Now, dear Jenny, I will answer the question you asked of me at home, whether I intended reenlisting in the Army. I can say Jenny that I shall not stay longer than my period of enlistment, which expires the 24th day of August 1864. Now that is plainly answered, is it not?

I am pleased to hear that you have engaged the school in your district, if you are bent on teaching the coming summer. It would be so lonely for your Mother to have you away. I received yours dated the 27th of March. You mention your father's being at home and his intending to get his leave of absence extended. I hope he may. 'Tis sweet to linger with friends. You ask me, Jenny, if my arm does not get weary writing. Yes, when I have much of it to do; but then I think it gains strength. I was very busy for nearly 3 weeks after I returned, but am not having much to do at present. My health is very good, and I think I shall soon feel like myself again.

I have not heard from home since leaving. I am in doubt in regard to the reason. In the first letter you said sister was unwell, and the second, said she was getting better.

'Tis getting late dear Jenny, and I must say goodnight. James.

April 13th. Yesterday 'twas cold and rainy, but today the bright sun is with us, and the gentle breezes that come to us remind us that the beautiful summer is at hand. I have just returned from the depot. Have been there for the purpose of drawing forage. I saw a company of Negro soldiers on the cars. Had a long

talk with the officer who commanded them. He said he never saw men so easily controlled and who would do their duty better than they.[3] Lt. Underwood
has gone to Chattanooga, and I am left to keep house. Billy Davis is in the
country, in charge of stock. I expect him today, and then I shall have company.
You said your mother had received a letter from Billy Bournes. We left him
at Chattanooga when we moved camp, and I have not heard from him since.
I have not seen Lemuel in a long time. I intend going to Chattanooga in a few
days, if I can procure a pass.

> Graysville, Georgia,
> April 21, 1864

Good evening dear Jenny; did you ever see a more glorious sunset? The
whole sky is an orange color, and the green treetops are tinged all over with
gold. There Jenny, do you see where the orb of day has disappeared behind
the rugged and rocky caps of Lookout range? I think I can hear you exclaim,
"Beautiful! Beautiful!" I was called away from writing a short time, say perhaps
half an hour, and what a strange darkness is covering the Earth with her dusky
mantle, and the campfires of our soldiers can be seen on the hillside and in the
valley in every direction. The bugles are singing throughout the camps, telling
us that roll call is at hand. I can hear the men answering as the orderly sergeants
call out their names. The candles are lit in the ("not the parlors") but in the
tents, and many like myself are writing letters to their dear friends; and why is
this? I will tell you Jenny; they think of loved ones far away, and think perhaps
that these dear ones are thinking of them. And that is the cause that prompts
them to write. Am I not right? Dear Jenny, oft the familiar words come to mind
and cause a bright gleam of sunshine to drive away the clouds that sometimes
cloud our hearts. "Do they miss me at home, do they miss me," and then we
fancy a voice saying, "Yes we miss thee at home, yes we miss thee; at morning
at noon and at night."[4] Jenny, is not imagination a dear blessing? Today I was
here in Georgia busily engaged at my work, and at the same time I was among
friends in the North, could almost fancy I saw their smiles of gladness and
heard their kind voices speak words of cheer. Only 4 months, and then if I see
fit, I can return to my childhood's early home, no longer a soldier, but free to
rove at will. You wished me to write and tell you whether I intended reenlisting
or not. I think I can answer you to a certainty, Jenny, that I shall not. My own
arm will not admit of another 3 years' trial. I think I have given you my views
on the subject in a former letter. I think you told me that your father's furlough
expired the 24th of this month, and then he was to leave his peaceful home for

the busy scenes of camp.[5] Ah how you all miss him, I know you do, but Jenny never despair. There is one above all others who cares for us, and watches over all, whether we be on the battlefield or enjoying the quiet of home.

The quartermaster has just come in and is singing, "I have roamed over mountains I have crossed over flood, I have traversed the wave rolling sand."[6] The other boys are around camp, making some friend a visit. Lemuel came to see us a few days since and stayed two days. He still remains with Lt. Drake. His health is not very good; I think he should be discharged.[7] I do not think a movement of this army will be made until the Potomac army is on the march. I do not think the day far distant when the main armies will strike, and that in conjunction. Friend Dixon reached camp the 22nd inst. He looks exceedingly well, and says he is rejoiced to get to camp again. He spoke of paying your folks a visit and having a good time. He said that my friends thought I was in a hurry to leave home. But Jenny this was not the case. Duty urged me to shorten my stay. They know not my heart if they think that home has no enticements to me. Goodnight. James.

Graysville, Georgia,
May 1, 1864

My dear Jenny,

I have just been reading one of Beecher's sermons, and now that I am through with its perusal, I will try and chronicle the events of the week.[8] There has been nothing unusual in a military point of view, only a few sharp skirmishes in the front. The enemy do not seem to have any great force this side of Dalton, a place on the Atlanta RR about 20 miles distant from here. Preparations have been going busily forward for an early move. All our surplus baggage has been sent to the rear, and the army put in trim for a long march. The weather is splendid, and the roads in good condition. The amount of luggage that is now allowed the officers and men is very limited. A half a shelter tent to a man and one of the same to an officer constitutes their habitations.[9] The transportation allowed to a regt. when we came into service was 14 teams. Now they talk of cutting it down to one. We have six teams at present, and I cannot see but what we get along as well as when we had the 14 and if they make it one, I am sure we can get along.[10] The soldier is a creature of circumstance, and is forced to take the world as it comes. I think, Jenny, the army a good school, if the lessons learned are rightly applied. Though I must confess, there are others that I should prefer a scholarship in.

What have you been doing all the week? Ah; I think I can guess. In the schoolroom, surrounded by a group of cheerful faces, laboring with all the zeal of a good teacher, teaching the young idea how to shoot. But hold, perhaps I am ahead of time. Your school may not have commenced. If not, then you have been enjoying the comforts of home and the dear society of a kind mother and loving brothers and sisters. Jenny, there are many who might envy your home and surroundings. Did you not know, Jenny, that grandeur of character was not always, or scarcely, ever found in the highest stations of life? 'Tis not rides on the farm that gives one comfort in this world. 'Tis in a nobleness of soul, and the pleasure of giving comfort to others, that true happiness is found.

The QM and several officers are talking over some of the incidents of our camp life during the last two years. To hear some of the scenes related, it seems as though we were living over the past. I have been invited out to a sing tonight, and where do you think it is? I will tell you. It is at the mansion of Messrs. Wells and Dixon. I think you will recognize the names as familiar. Ask Darius if he will accompany me, we need an alto singer. I think he would enjoy it very much. We have many pleasures in camp, and are as a general thing a happy set. Why be otherwise? Take life as it comes and make the best of it, is a good motto.

The time of our enlistment is drawing rapidly to a close. May, June, July, and a part of August. The time will pass rapidly when we get to marching. 'Tis not near as irksome and monotonous as lying in camp. When passing through the country, there are many things new and exciting to call our attention, and time flies swiftly.

I am in hopes that they will call the militia of the Northern states into the field for the summer's campaign. I think this the best thing that could be done to crush the rebellion. We have hundreds of troops garrisoning the forts in our rear that could [be] brought forward, and their places supplied with new levies. I see that the governors of Ohio, Ind., and Illinois have already taken action in the matter, and 'tis a timely move.[11] The coming campaign ought to be decisive. If not, then the war will be prolonged. Let it not be for want of men that we fail. I say fail, we shall not fail so long as there are men enough to make an army.

I received your letter dated the 17th of last month, with the early flowers of spring enclosed. I did not think the blossoms of spring were with you yet, as you wrote me that the season was very backward. They do remind one of the days of childhood. Our early days are the bright sunshine of our lives, and the first flowers of spring are sweetest. How I should have enjoyed a ramble with you. The bugles are blowing for lights out, and I must bring my long letter to a close. You say you think Dixon's stay at home was not so irksome as some would think. I do not know, I am sure. He often speaks of it as a pleasure. The boys tell that it is getting late, and wish me to bring my letter to a close. I wish you a kind goodnight. James.

Buzzard Roost Gap. (Guernsey and Alden, *Harper's Pictorial History of the Great Rebellion*)

May 2nd. There is a rumor in camp that our division moves to Ringgold to-morrow morning. It is six miles south of here on the Atlanta RR.

George Thomas's 14th Corps rendezvoused around Ringgold, and Grant and Sherman soon launched their campaigns against the armies of Robert E. Lee and Joseph E. Johnston, Braxton Bragg's successor. The familiar call of the bugle prodded the Michiganders to consciousness early on the morning of May 7. The Army of the Cumberland—one of three armies cooperating under Sherman's overall command—was set to advance against the Army of Tennessee. Brigadier General John Haskell King was in command of the 11th Michigan's brigade.[12]

Sherman advanced toward Atlanta in a series of successful turning maneuvers around Johnston's left. The 11th Michigan played its part, coming under fire at Buzzard Roost Gap on May 8 and at Resaca on May 14. The Michiganders then pursued the Rebels to Cassville, where the Confederates were again dislodged, this time with a punishing artillery barrage. Sherman next turned Johnston out of Allatoona Pass, but the Rebels shifted quickly enough to bar the Federal advance on a line running from Dallas to New Hope Church. The 11th's brigade again came under Colonel Stoughton's command, due to a wound incurred by division commander Richard Johnson. The regiment held its position under fire through June 5, tasting the misery of trench warfare. Sherman gradually extended his left back to the Western and Atlantic Railroad, and then resumed his advance despite persistent rains. The Confederates assumed a strong defensive position at Kennesaw Mountain on the night of June 18, and a thunderous Federal bombardment ensued the next day. Days of intense skirmishing followed, culminating in a disastrous Union frontal assault against the heights on June 27. The Wolverines were not engaged that day, but the men anticipated orders to charge the formidable Rebel works at their front and passed the hours in excruciating suspense. The regiment was relieved from the front line after the attack and relocated to a wooded area half a mile back.[13]

Four miles south of Kingston,
Georgia, May 21, 1864

My dear Jenny,

Since leaving Ringgold, we have marched a long way and have had some heavy fighting. Our regt. have had one killed and 2 wounded. At Resaca the fighting was heaviest. The 25 Mich. suffered severely, so I am told, and also the 19th Mich.[14]

The boys are all well, and in fine spirits.

I cannot write you a long letter, but when I have an opportunity, I will give you the particulars of our campaign. Remember me kindly to your mother and the rest. James.

I rec'd your letter dated May 1st. Be cheerful Jenny, there is a better day coming.

Resaca, May 14, 1864. (*Frank Leslie's Illustrated History of the Civil War*)

Near New Hope Church,
Georgia,
May 29, 1864

My dear Jenny,

We followed the enemy from Kingston, and he is again in position giving us battle. There has been severe fighting, and the thing is not yet decided. This battle determines the fate of Atlanta. We have had 5 wounded in the regt. up to this writing, among the number is Dixon. A shell burst over his head and one of the pieces hit him, making a bruise, but nothing serious. I do not apprehend.[15]

I have rec'd 2 letters from you since leaving Kingston, for which I am truly thankful.

Excuse short letters. Please write often. James.

Camp, south of Big Shanty,
Georgia, June 15, 1864

My dear Jenny,

Seated at the roots of a large oak tree behind the breastwork, I am penning you a few lines. 'Tis a pleasant morning, and one could not think that the bloody strife was going on, did he not hear the thunder of cannon, the whizzing and bursting of shell near at hand. I am feeling well this morning, and most of the boys are in good spirits. We are gaining ground slowly. The enemy seem determined to contest every inch of ground between here and Atlanta.

Excuse briefness Jenny and remember me as ever your James.

On the RR 4 Miles from Marietta,
Georgia, June 20, 1864

My dear Jenny,

Our work is progressing favorably. There was heavy fighting yesterday in front of Kennesaw Mountain. We gained a decided advantage, driving the enemy nearly a mile, causing him to evacuate his strong works and fall back to the mountain. Our division was in reserve yesterday, but we could see the enemy in line of battle clearly from where we lay. Skirmishing and artillery firing is going on this morning. We have had one wounded in the regt. since my last writing. I think that we shall have possession of Atlanta ere the 4th of July.

The cars come in sight of us, bringing in supplies and rations.

Your letter of the 7th inst. is at hand, came to me yesterday while watching the angry and sullen surges of battle. "Mid the smoke of the contest, the cannons' deep roar" it found me, and Jenny, how can I thank you for the kind words which

Kennesaw Mountain, June 22, 1864. (*Frank Leslie's Illustrated History of the Civil War*)

it contained?[16] Need I say that it carried me back from these scenes of strife and turmoil to the peaceful surroundings of home, friends, and loved ones?

A letter from Dixon places him in [the] hospital at Nashville.

The arm gets no worse, and I think in time will assume its naturalness. Tell your mother I thank her kindly for her charming words. Write me often of your father's whereabouts.[17] May kind heaven bless and protect you Jenny. I am as ever your James.

> Near the foot of Kennesaw Mountain,
> Georgia, June 24, 1864

My dear Jenny,

There has been continuous fighting since my last writing.[18] We had one man killed yesterday and one wounded. There was continuous fighting along the whole line, both musketry and artillery.

My brother John was with me nearly all day yesterday. He has escaped the fiery ordeal unscathed.[19]

I rec'd your letter of June 1st yesterday. You complain of my not writing often. Now Jenny I think I am perfectly excusable. Only think, we have been marching and on the battlefield since the 7th of May.

There is not much doing this evening, only skirmishing. Everything is working favorable, and I think ere long the Confederates will acknowledge another defeat. This war cannot always last. Even now, I think I can see the final end. Write me often, and of your father and his whereabouts. Don't forget me to your mother, and remember one who will ever prove true. James.

> Kennesaw Mountain,
> Georgia, June 25, 1864

My dear Jenny,

Your letter of the 13th inst. is at hand. I think the explanation I have given you in my last, stating the reason why I did not write, will be satisfactory. You think I am making a trial of your faith. Far from it Jenny, and you do me great injustice to let such a thought enter your mind.[20]

There is nothing new in our front since my writing, but the wings of the army have gained advantages. Enclosed I send you a sketch of the mountain, showing the enemy's position.[21] Our army encircles one end and two sides of it. 'Tis getting so dark, I can barely see a mark. May heaven bless you, dear Jenny, is the earnest prayer of James.

> Kennesaw Mountain,
> Georgia, July 1, 1864

My dear Jenny,

Everything is working well. The enemy's position is a strong one, and it must take time. I think there has been one of our regt. mortally wounded since my last writing. His name was Wm. Mansion, Co. G.[22]

Things have been very quiet in our immediate front for the last two days. I am sure, Jenny, if you could see me and the advantages I have for writing, you would not blame me for short letters. This is the first letter I have written with a pen since the commencement of this campaign.

My account book serves me for a writing desk. My arm seems to gain strength, and it does not pain me as it used to. This is a glorious day. The air is cool and refreshing. The regt. is camped in a shady grove. We have been here 3 days. The boys are getting pretty well rested. One of our boys rec'd a letter from a member of the 6th Michigan saying that muskets had been given them and that they were 30 miles from Port Hudson.[23] How does Lemuel get along, and does he depict in glowing colors the beauties of soldiering? Jenny in one of your letters, you seem to have had some difficulty in your labors as a teacher. I hope nothing serious. You say, "Would that I knew my duties as a teacher." Now I am sure that you have done your duty as well and ably as you know how, and when one has done his or her best, angels could do no more. I think Jenny that when you have done your best, you could not ask to do better, nor it should give you bitter thoughts. Let this console you, that you done your best. Perfection is seldom ever gained. Then Jenny if we use our best endeavors to gain a certain point and fail, ought we to grieve? Are we to blame? I think not.[24]

My hand is getting wearied, and I must close. Goodbye dear Jenny, for this time. James.

On July 2 the Rebels were yet again turned out of their defenses, and Stoughton's brigade skirmished sharply with the Confederate rear guard the next day. James was accompanying the colonel a short distance behind division commander King when the general and his staff—oblivious that they were the lead element of the Federal army—were fired on by Confederate cavalrymen, who promptly fled. George H. Thomas soon arrived. "You are in the advance," Thomas told General King as James listened in. "Throw out two or three companies of the 11th Michigan as skirmishers, and continue to push right along as you have been doing. . . . If the Rebels cause you too much trouble order up some of the artillery and *scrawl* the canister to them [emphasis in original]." The chase continued, and the 11th Michigan fought in the Battle of Ruff's Station on July 4. James then left the regiment for three weeks to watch over Colonel Stoughton, who was dangerously wounded in the fight. The Confederates in the meantime retreated across the Chattahoochee River, and Sherman's move against Atlanta itself commenced on July 17. Johnston was replaced with John Bell Hood, who immediately lashed out at the Federals at Peachtree Creek. On July 22, the 11th's brigade advanced and entrenched within two miles of Atlanta. Holding position, they were excluded from the Battle of Atlanta that day and from the subsequent clash at Ezra Church on July 28.[25]

John Haskell King.
(Library of Congress)

Marietta, Georgia,
July 7, 1864

My dear Jenny,

It is several days since I have written you. I think we were then in front of Kennesaw Mountain. The morning of the 3rd of July, it was ascertained that the enemy had played one of his old tricks; that is, retreated under cover of darkness. At daylight, the whole army was in motion and in hot pursuit. Nine o'clock found us in Marietta, the pleasantest little village I have seen in all the sunny South. The rear guard of the enemy had just left where we entered. We pursued them some 3 miles farther before finding them in force. We then pushed them steadily about 4 miles, when it was ascertained that they were strongly entrenched, behind good fortifications. Here a halt was ordered, and our men commenced fortifying to put them on an equal footing with the enemy. During the night our men worked diligently, planting batteries to command the enemy's works. The morning of the 4th of July dawned clear and pleasant, and was ushered in by the discharge of cannon from both sides. Shells went flying through the air, making a whizzing noise that could be heard a mile or two, then the explosion and bursting into a thousand fragments. The morning passed swiftly away. Ten o'clock came, and with it, orders to charge the enemy's rifle pits & carry them at the point of the bayonet. Only 3 companies of our regt. was in this charge. They done the work handsomely, but suffered a loss of 3 killed and eleven wounded.[26] Col. Stoughton, commanding the brigade, was struck by a piece of a shell and right leg broken, shivered in so shocking a manner that amputation was necessary. The next day after he was wounded they carried him to Marietta, and he requested that I would go with him. I shall remain until his wound is better. We have a fine brick house in a beautiful grove, and everything convenient.

I saw my brother John the day I left the regt. He was enjoying good health, and in good spirits. I cannot write more now, write often and direct as you previously have. I remain as ever yours. James.

Marietta, Georgia,
July 16, 1864

My dear Jenny,

Last night I rec'd your letter written July 1st, and you still complain that I do not write. Truly Jenny, for the last few weeks I have written you so often that I was fearful I should flood you with correspondence. 'Tis true that from the time of leaving Ringgold I did not write you, the reason, I had no opportunity; but lately I thought I was doing uncommon well. I am in hopes that the next I receive from you will acknowledge the receipt of some of the many I have written you. I think when in front of Kennesaw Mountain I wrote you no less than 3 times, and in one I sent you a sketch of the mountain; all of which I have no acknowledgement of their being rec'd. Jenny the twenty-fourth of August is hastening around, and soon the time will come when we shall be mustered out of service, and allowed to enjoy the society of friends and comforts of home from which we have been so long separated, and denied. Oh, will it not be a "joyful day" to the one who has passed through the dangers and perils of a life of 3 years in the service of one's country? You think sometimes that I intend staying longer than 3 years in the service. Give yourself no fears. I think I have done my duty, and further than that I will not go. Would this "cruel war" was over, and all soldiers permitted to return to their homes. Then indeed, we could all rejoice. There has been none of the regiment injured since the 4th of this month.

Some of the boys come in from the front nearly every day. Colonel Stoughton improves slowly and the chances are that he will recover. The Lord grant that he may. The regiment seems as though it had lost its father. Oh Jenny how I wish you could visit this beautiful place. The streets and buildings are completely embowered with trees of the most magnificent foliage. The former inhabitants showed their good sense by leaving many of the forest trees. Our yard is filled with oak and hickory, besides many of the kinds of trees which grow only in a Southern clime. In the corner of the garden stands a fig tree, loaded with green fruit. The former proprietor made up his mind to let the Yanks eat of his fruit, and he has gone in quest of his rights. The weather is decidedly warm, and I must say very uncomfortable, but the greatest pests I have are the flies who cause me to lift my pen ofttimes in the middle of a word to brush them from my face. Would that they had gone south with their owners. I have drawn the blinds so closely I can hardly see to write, but I think they can see to do their mischief equally as well as I can see to transcribe. In one of your letters you said that it seemed as though I "were trying to test you." I wrote you explaining all, but it seems you did not get it, and I cannot let it pass without saying a word in regard to the matter. In the first place Jenny, why should I test; what reason have I for doing it? Why do I test her who has been already tested? Testing one in whom I have the utmost confidence, and never for a moment have doubted. Ah; Jenny you wronged me. And I am sure, if you had taken the sober second thought you would not have written this. You do not know the pain it gave me, but when I thought the matter over, I knew that if you thought so, it was only for a moment. Hoping that you have received my letters ere this, and everything satisfactorily explained, I remain, as ever, yours, James.

> Marietta, Georgia,
> July 22, 1864

My dear Jenny,

'Tis a rainy, damp, and disagreeable evening. Just the kind of a night to make one melancholy, or have the blues; yet I cannot say as I am affected either way. The rain comes in fitful gusts. I can hear it patter in the branches of the trees near my room. I do not hear it on the roof, as we used to sing, for the simple reason that I am in the story nearest the ground. Our song you remember, "Every tinkle on the shingles, has an echo in the heart, and a thousand dreamy fancies, into busy being start." Ah, those dear bygones. I am watching with the colonel tonight. He is much better than when I last wrote, and there is not much doubt but that he will be spared to return to his home and family. The first few days his life was despaired of, but now there is not much danger. His father-in-law, Mr. Page of Sturgis, is here and assists in caring for him, which is a great help to us.[27]

There is a rumor that Atlanta is taken.[28] I have heard no particulars whether the fighting was very severe or not; but that our troops occupy the place I have not the least doubt. I can hear cannon in that direction this evening, which tells me that work is going on somewhere. This has been a long and tedious campaign. Only think we left Ringgold the 7th day of May, fighting commenced and is still going on. I had supposed when we started that Atlanta was the objective point of the campaign; and the capturing of that point would end the controversy in this quarter. Whether General Sherman will go into camp and rest the army or keep fighting the enemy, remains to be seen.[29] I think fight is his motto. Perhaps our campaign will not end until things assume a quietness on the Potomac, and around Richmond. Everything is working well for the Union cause. We are meeting with successes every day. 'Tis true, Washington has been threatened and the Rebels threatened a Northern invasion, but that danger has passed without even disturbing the equilibrium of Mr. Grant, who still continues to peg away at Petersburg.[30] I have no doubt that the enemy thought, in reaching a force northward to threaten Washington, it would cause Grant to withdraw his army, but in this they utterly failed, and I am in hopes before many days we can say of Richmond, as we now do of Atlanta, that it has fallen.

Write soon and let me know whether your father has joined the regt. or not. I wrote him a few days ago, and directed to New Orleans.[31] Don't fail to remember me to your mother. I remain as ever your James.

> Two Miles from Atlanta, Georgia,
> July 26, 1864

My dear Jenny,

I reached the front yesterday afternoon and found the regiment behind good fortifications, taking the world as easy as circumstances would permit. The rumor that Atlanta was taken, and I think I wrote you that such was the case (while at Marietta), proved false. Our shell fall into the city. We have taken a number of prisoners. I met 1,240 going to the rear yesterday. How long they may hold out I cannot tell; but I am in hopes [that] when I write you I can chronicle its downfall. I cannot write a long letter. Excuse haste dear Jenny, and believe me as ever your James.

Near Atlanta, Georgia,
July 30, 1864

My dear Jenny,

There was a heavy battle fought to our right day before yesterday, resulting in the defeat of the enemy. The day before, the 15th, 16th, and 17th corps, which had been posted on our left, were transferred to the right wing in order to take possession of [the] Macon and Atlanta RR. The 15th Corps had taken position and were getting very good works erected when they came on them in heavy numbers, charging in the most desperate manner. Our men repulsed them. They then, gathering fresh numbers, charged again, only to be repulsed. The musketry firing was terrific. One of the 16th Corps who passed over the field said that the enemy were literally piled up. Their loss was much greater than ours, owing to the fact that we fought behind breastworks.[32]

I think there has been two of the regt. injured since my last, though not seriously. From a hill near our camp can be seen the City of Atlanta. Among the most prominent buildings can be seen several churches, a gin distillery, and the courthouse from which the Rebel flag now flaunts. There is a report that the enemy are moving their stores and preparing to evacuate. I am in hopes that it is so.

The weather is very hot. It has the appearance of rain this afternoon. There has been very heavy cannonading in our front today. They have thrown several shots near our quarters this afternoon, but no one has been injured as yet.

I cannot write more this afternoon, 'tis too warm. Give my best wishes to your mother, and remember me as ever your James.

––––––––

Previously, on the night of July 26, James's childhood friend Charles Rice had eaten his evening meal with the brigade band half a mile back from the trenches. As he and his fellow musicians began preparing for the night, a stray Minié ball pierced Rice's back, passing beneath his shoulder blade and penetrating his lungs. His comrades feared they would lose him at any moment. Upon hearing the news, James rushed to his friend's side. The surgeon proclaimed the wound mortal. Charles smiled at the sight of his old friend, but when he tried to speak, his mouth filled with blood. An ambulance carried Rice off to Vinings Station the next day, and a hospital train transported him to the general hospital in Kingston.[33]

Perhaps James was mulling over the randomness of Charles Rice's fate when he commented to Jenny, "They have thrown several shots near our quarters this afternoon, but no one has been injured as yet." Later on July 30, after mailing his letter, James found a pleasant patch of grass and sat down to relax with Lindley Harkness. Billy Iddings and John Underwood were present as well. They were situated well back from the battle line. James buried his nose in a good book, *Country Living and Country Thinking* by Gail Hamilton. A shell issued forth from a distant Confederate battery, arced up, reached its zenith, and assumed an alarming trajectory upon descent. Before anyone could react, the missile burst in the air overhead. A large fragment struck James on the back of his left shoulder, knocking him flat on the ground, unconscious. Harkness lifted his wounded comrade, set him on some blankets, and yelled for Iddings to fetch Doctor Elliott.

When James regained consciousness—more than half an hour later—Elliott was tending to his broken left shoulder. The joint was throbbing and severely inflamed. The patient was soon removed to the field hospital one mile to the rear, where he spent two days before being evacuated to Vinings Station. At Vinings he was treated by the 11th Michigan's Isaac Kemberling. James knew Kemberling well; the same man had dressed his Missionary Ridge wound for weeks at the hospital in Chattanooga. On August 6 James wrote to Jenny, "I am very comfortable today. My shoulder and arm are much swollen, but are not as painful as they have been. I think I shall be able to come home with the regiment."[34]

James recovered sufficiently that he was permitted on August 17 to lodge with Addison Drake in Chattanooga. One day shortly thereafter, James felt an urge to visit the railroad depot. As he strolled along the depot platform, a freight car pulled into the station. He was astonished as "a skeleton of a soldier came tottering from that car and said, 'Jim, I knew if I could find you, I would be taken care of.'" It was Charles Rice. James recorded his friend's macabre adventure:

"Why, Charley, where did you come from?" [I replied.] The answer was, "I ran away from that hell of a general hospital at Kingston." His story in brief was this: The night before, about nine o'clock, he overheard a conversation between two of the assistant surgeons, and one of them, alluding to him by name, said, "He will not live until morning, and I am going to cut him open and find out just the course of that ball, and the effect it had on the lungs." The hospital tent was close to the railroad platform, and the wounded and almost dying comrade, summoning all the strength and willpower at his command, rolled himself from under the flap of the tent, onto the railroad platform, and from there into a boxcar, and soon it was coupled onto a train and fortunately brought to Chattanooga. He was so feeble that he could stand but a moment without support. [I] borrowed a couple of blankets from soldiers nearby, procured an ambulance and carried him to the quarters of [my] officer friend, Quartermaster A. T. Drake. Here he remained three or four days, and then [I] accompanied him to the hospital where [I] had formerly been an inmate. [I] made full explanation in regard to [my] comrade's desertion from the Kingston hospital, and the surgeon said he would straighten matters out so that the sick and wounded comrade should not be carried on the rolls as a deserter.... Under the skillful and kind treatment of the surgeons in charge of the Chattanooga hospital, our soldier musician in a short time was able to be taken north to Jeffersonville, where he served in the invalid corps to the close of the war.

Charles Rice would outlive his brush with death by almost forty years.[35]

———————

Chattanooga, Tennessee,
August 18, 1864

My dear Jenny,

I was moved from the field hospital at Vinings Bridge the 11th of this month, and reached Chattanooga the afternoon of the 12th. The journey was very fatiguing. I was placed in [the] general hospital at this place, where I remained until yesterday, when Lt. Drake obtained permission from the surgeon in charge to take me to this office. 'Tis much pleasanter here among friends than in the hospital. You know Jenny my description of hospital life was not very glowing if you remember in regard to my first initiation.[36] My arm does not give me much pain, and I think a few weeks will mend it as well as ever. How strange it seems to me that I am once again in this city where so many incidents have occurred in my soldier's life. Well do I recollect my first admit into the place. 'Twas a clear, cool September morning when our long train of army wagons came winding along the Chickamauga valley road and camped at the foot of Lookout, on the banks of the blue Tennessee. Far away to the south could be heard the fearful throb of battle, borne to us by every passing breeze. The next day we learned that our army had been overpowered, but not vanquished, and were falling back on Chattanooga. There followed long days and weeks of suffering on account of the scarcity of provisions. Oh how joyfully we hailed the dear old flag the morning of the 24th of November when borne by the veterans of Hooker. We saw the enemy driven around the northern end of Lookout like chaff before the wind. Then the brilliant charge of Mission Ridge, and the days and nights of suffering on account of the wound I received. From here I started on my journey home to see relatives and friends after an absence of nearly two and a half years. After a short stay at home, I was once more among the scenes of this noted city, and now after passing through the most remarkable campaign the world ever saw, I return to its limits to gaze on scenes long familiar.

I understand that the regt. has been relieved from the front and are now at Marietta.[37]

I cannot write more, dear Jenny, and am in hopes that I shall see you in many days. James.

————

The 11th Michigan Infantry was relieved from the front on August 27 and arrived back in Sturgis, Michigan, at dawn on September 25, 1864. Discharge took place on October 13. Of the 1,004 men who departed Michigan in December 1861 with the unit, only 340 were present to muster out together. All of the officers present, including William Stoughton, signed a statement penned on the back of James King's discharge paper. "The within named, James W. King, was distinguished as a soldier for the ability and zeal with which he performed every duty pertaining to his office, as well as for the integrity of his character and his gentlemanly deportment. Taking a voluntary part in the battle of Mission Ridge, he behaved with conspicuous gallantry, and received a severe wound, also another before Atlanta, in August 1864. We cordially recommend him to the favorable consideration of all civil and military authorities."[38]

For many of the soldiers, the biggest battle lay yet ahead: reintegrating themselves into a forever-changed civilian world. The war had disabled James King; no more could he till the soil of his father's farm. In mid-January 1865 he set out for the familiar state of Tennessee, leaving Jenny behind in Michigan for the time being. The ex-soldier sought a means of quick financial success in the South, and he soon secured an annual salary of $1,500 working as a clerk with Billy Iddings under their old friend and comrade Daniel Webster Holbrook, who was now chief clerk at the Chattanooga office of Assistant Quartermaster Ezra Benham Kirk.[39]

There James witnessed the Confederacy's death throes from a safe distance. On April 10 he wrote to Jenny from his office:

This has been a day of rejoicing with us. About 10 o'clock a dispatch was received that Gen. Lee had surrendered his army to Gen. Grant, and since that time cannon have been firing, engines whistling, soldiers cheering, and flags are floating in every part of the city. We have two large flags run

Silk national flag of the 11th Michigan. (Photo by the author)

out of the front windows of the office and a large one on the wall inside, and we have made preparations to have a grand illumination. This will be a day of general rejoicing through the Union, and I do not suppose our celebration here will hardly be a drop in the cup, as the same thing will be done in all parts of the land where the magnetic wires extend. 'Tis not the great amt. of men, guns, & c. captured that causes us to rejoice, but the prospect of such an event bringing a speedy peace.

I stopped my writing to see the illumination. The whole city has been in a blaze of light. Soldiers throng the streets cheering and shouting, all seem intoxicated with the cheering news, and some who have partaken of the ardent [spirits] are quite so.[40]

His next letter, penned six days later, reflected the stunning change of national fortune, and an accompanying loss of goodwill toward the conquered.

The news of the assassination of President Lincoln came to us by telegraph yesterday, and turned our day of rejoicing into sadness and gloom. At first we did not credit it, but dispatch after dispatch came until it could be doubted no longer, and the deepest sadness pervaded all circles. Flags were lowered to half mast, others suspended from the fronts of buildings clothed in mourning. All places of business were closed, and on every door was seen a piece of crape, and all day today from sunrise until sunset at every half hour, a gun has been fired. It came so unexpected to all, "like a clap of thunder in a clear sky." Had it been four years ago when it was threatened, we should not have been so shocked, but at this late day, when the gigantic rebellion has almost ceased to breathe, and the deluge of blood which has covered the land for the last four years nearly passed away, to have him who has stood at the head and guided us safely through the storm stricken down by the hand of an assassin, it is the worst calamity that could have befallen the nation. The assassinator I saw on the Louisville stage in 1861. What should have led him to commit the rash act is more than I can say. All his reputation he owes to the North and [to] all his kindred. The name of Booth has sunk forever into ignominy. May he and his companion in crime meet the fate they really deserve. When the news came yesterday, one of our employees cried, "Good, it ought to have happened long ago." He was immediately arrested and sentenced to three years' imprisonment in the Alton penitentiary. Several others for expressing a gladness that such was the case are now wearing chains and balls of iron as ornaments. Hanging would be too good for these villains, had I their sentencing.[41]

Kirk's quartermaster office was transferred to Nashville soon after. James spent much of the balance of 1865 traveling the length of the Mississippi River to inspect maritime wreckage from the war, and to determine the feasibility of salvage operations. His itineraries included Cincinnati, St. Louis, Vicksburg, and New Orleans, and he stayed in some of the country's finest hotels, such as the Lindell in Cincinnati and Burnett House of St. Louis. The young Michigander was impressed with the urban sights, but he soon wrote to Jenny, "The more I see of the city life and people, the more disgusted I am with their ways and manners."[42] He remained a small town boy at heart.

James returned to Michigan in the fall. On October 6, 1865, James Wood King and Sarah Jane Babcock were joined in marriage by Reverend Luther Trowbridge. James returned south without his bride for now, but his months alone in Dixie were drawing to a close.[43]

Many familiar faces from the 11th Michigan also headed south again, lured by the promise of easy money. Melvin Mudge, William Elliott, Billy (Will) Iddings, and Linus Squire all returned to Dixie in hopes of buying or leasing cotton plantations for pennies on the dollar.[44] Emancipation, cemented by the

Chattanooga Tennessee
April 16th 1865

My Dear Jenny,

I should have written you ere this but Communication has been again interrupted The heavy rains have washed away bridges and sent them whirling down the streams and the consequences are that we have had no trains since last Monday. The road is nearly repaired and a train is Expected in tomorrow morning I am sure that it will bring me a letter from you if it does not I shall be disappointed But then I shall not let it grieve me for I know twill not be many times that I am doomed of Jenny I can write.

The news of the assassination of President Lincoln came to usby

James's letter to Jenny regarding the Lincoln assassination. (Courtesy Western Michigan University Archives and Regional History Collections)

REGISTER OF HUSBAND.

Name *James W. King*

Birthplace *Three Rivers, Mich.*

Date of Birth *August 13, 1842*

Descent *French and English*

Father's Name *Benjamin W. King*

Mother's Name *Martha Weatherbee*

No. Br's & Sisters *Seven — 3 Sisters, 4 Brothers*

Education *Common School*

Occupation *Editor*

Politics *Republican. Cast first vote for Abraham Lincoln in 1864*

Religion *Free Thinker*

Date of Marriage *October 6, 1865*

James W. King's wedding register. (Author's collection)

Thirteenth Amendment, had obliterated the existing system of capital in the South—slaves had long served as leverage for credit—and many plantation owners, left with no means to pay for agricultural labor, necessarily resorted to leasing or selling their land. James was no more immune to this temptation than his comrades. One of these men, Linus Truman Squire, had charged up Missionary Ridge shoulder to shoulder with James until being disabled two-thirds of the way up the slope. Now King and Squire determined to face the new year of 1866 confronting another risky, but altogether different, venture together: they leased the Critz cotton plantation in Thompson's Station, a small town in Williamson County, Tennessee.

CHAPTER 7

King Cotton

December 1865–January 1868

—⬩⬩⬩—

I expect to make this land of the South
my home for some time to come.

The bitterness of Reconstruction Era politics would ultimately spawn the carpetbagger stereotype: that of a morally and financially bankrupt individual, drawn from the lowest element of Northern society, who invaded the South as a predator to feed off the suffering of the conquered, to seize and abuse political office, and to raise the supposedly childlike, inferior freedmen above their rightful station. But in fact the Northerners who came south, on the whole, were educated, well-heeled citizens. The vast majority were successful businessmen and professionals, and most were Union army veterans. They came to teach; to serve as lawyers for the freedmen; to become agents in the Freedmen's Bureau; to publish newspapers; to sell wares; to farm land that would otherwise have deteriorated from neglect; and to invest large sums of desperately needed cash into a devastated economy. Most did not come seeking any role in Southern politics. Those who came to run cotton plantations did seek personal success and financial reward, yet this was the very same impetus driving America's great westerly migrations, and pioneers are remembered in an overwhelmingly positive light. At the time, the Northern transplants were seen for what they were, and those who arrived in 1865 and 1866 typically met with polite greetings.[1] Thompson's Station was no exception. In fact, Will Iddings had already settled there with his wife-to-be, Fie Johnson.

The family of Jacob Critz Sr. had arrived in Thompson's Station in the 1820s, and built a family home there circa 1835.[2] As of 1860, Critz's son, Jacob Critz Jr., employed a dozen slaves, and in 1866 he retained the same three hundred acres of fertile soil he had possessed before the war. James and Linus's field hands may have included some of the Critzes' ex-slaves, who had probably first tasted freedom just one year before.[3] The names of James and Linus's freedmen field hands and domestic servants are gradually revealed through correspondence, and appear to have included Harriet, Aunty, Manda, Kate, Crockett, Melvina, Jimmy, Harold, Robert, Martha, and Josh. Two more, Lawson and Julia, were evidently children.

James and Jenny were together in Thompson's Station from late January through early September. That period is sparsely documented, illuminated only by letters that passed between the young couple and their parents. James wrote the first letter below while visiting a plantation near Murfreesboro that Linus Squire had leased for the 1865 crop season.[4]

———————

Winstead Farm,
near Murfreesboro, Tennessee,
December 28, 1865

My dear Jenny,

This evening finds me seated by Squire's hospitable fire, talking over the chances for a cotton crop.[5] Left Nashville at 3 o'clock yesterday afternoon but did not get here until after dark, and found Squire leaning up against the telegraph pole, looking anxiously for me. We have made up our minds to rent in Tennessee, and shall go to Pulaski tomorrow to look up a farm and rent if we can. The season is getting so far advanced that if we should take time to go to Mississippi or farther south, it would be very late before we could get settled. So make up your mind to come to Tennessee.

Today we have been to Murfreesboro and had several talks with cotton men. Some think we had better go farther south, while others think the next year we had better stay in Tennessee.

Shall be glad when we get settled and you are with me. I never used to get homesick. Don't think I am now, but then there is someone I want with me, and you know who that is.

Should like to see you all. Does Ettie make as much noise as usual?[6]

Write me a long letter, and direct to Murfreesboro, Box No. 173. Will have you come as soon as we can get things arranged. Your James.

Linus Truman Squire. (U.S. Army Heritage and Education Center)

Nashville, Tennessee,
January 3, 1866

My dear Jenny,

I have just returned from my trip southward. We succeeded in renting a farm between Franklin and Columbia, and you can get ready to start at an early day. We rented 200 acres of good cotton land with a fine brick house upon it and everything comfortable, within 2 miles of the R.R. Pay $5.00 pr. acre.

Send the box as soon as you get it ready, and mark it to Nashville in care of Capt. E. B. Kirk, AQM. Purchase whatever you deem necessary. If father does not get more than $50 for me, you can use 30 or forty in purchasing, and then have enough left for traveling expenses. Will write you again and tell you when to start. Remember me to father, mother, brothers, & sisters. Your James. You may start Wednesday if you are prepared. Father can give you the proper directions in getting your trunk checked. At Nashville, if I am not here, get your trunk checked to Franklin Depot. If you arrive in the evening, you will have to lay over here one night, then come to Thompson's station. Will meet you here if it be a possible thing. Come by way of Salem Crossing and Indianapolis.

Winstead Farm,
near Murfreesboro, Tennessee,
January 5, 1866

Dear Jenny,

This morning while at Nashville, I mailed you a letter and I did not have much time to write, so now I will finish what I began. Such a tramp as we have had finding a location to suit. We left Nashville Saturday at 7 o'clock intending to go to Pulaski, but when we got to Lynnville we found two bridges washed away and we could go no farther, so we got off in the woods and held a consultation. A gentleman on the train told us we would find good cotton lands about Cornersville, about 12 miles east of where we then were. So off we started on foot, when did I ever see such muddy roads? We got to town just as the sun was going down. Upon inquiry, we found that all cotton lands had been rented in that vicinity. So all we had to do was to make our way back to Lynnville, which we did Sunday morning. Came back as far as Thompson's Station, and after 3 days' hard work, we made a choice.

Write me when you can start. Shall look for a letter soon. Your James. This is wretched paper, a wretched pen, and the ink closely related to both.

Nashville, Tennessee,
January 8, 1866

My dear Jenny,

I sent you a telegram this morning not to start until Monday next. It will take longer to move than I expected. Think we shall get settled this week. I shall look for you Wednesday morning at the Nashville Depot. If either of our fathers will come with you that far, I will pay all expense, though I do not expect you would have any trouble.

I went down to Thompson's Station yesterday and made out writings for the farm, paid $500 down, and the other $500 I pay the 31st of next month. Shall not have much furniture to buy, as the man I rent of leaves most of his in the house. We pay $5 pr. acre and intend [on] hiring Negroes by the month. Most of the planters are renting lands at 6 or 7 dollars pr. acre, find all stock and implements, and give the Negroes half.[7] I think if they make anything, we ought to, surely. Don't fail to write. Your James.

Ask father if I did not leave money with him to pay the taxes on my land the last time I was at home, I think it was $20.

Thompson's Station, Tennessee,
January 26, 1866

Dear friends at home,

Jenny arrived at Nashville the evening of the 24th, having made, as she termed it, "a successful voyage." She was somewhat wearied with her long ride, and we remained at Nashville until yesterday evening, when we took [the] train for Thompson's, and 5 o'clock found us safe at home at the farm.

Wish you could drop in this morning and see how well we are situated in our new home. Jenny thinks she "shan't get homesick one bit." Can only write a few lines this morning, Jenny will write in a few days and tell you all the news. James.

[To Jenny from her mother, Ruth Butler (Blodgett) Babcock.—Ed.]

Fabius, Michigan,
March 11, 1866

Dear child,

We received your very welcome letter mailed March 1, glad to hear you was all getting along so well. I was pleased to hear you was going to have one Northern family for neighbors. Hope you will enjoy their society. I read their marriage in the Reporter.[8] You have been gone a long time. Does the time seem long, or short to you? Have you seen any other white woman since you got to your home? Now tell me, Jenny, what that Southern climate does to change people so soon. I must confess my stomach felt riled when I heard you kissed a—I always thought you had a great dislike to such things. Don't blame Jimmy one bit if he did not like it. Now if he will only bring you back here I am sure with the aid of the climate I can cure you of all such heathenish tricks. Now James said you was not homesick one bit. Can he tell me so now? I don't expect you to ever be willing to leave Jimmy for the rest of us, but I can't help think my little Jenny would like to see us by this time, but you must be a good girl, and I think some future day we shall have that privilege. Oh Jenny how many happy hours we have spent together. I can never forget them, and my earnest prayer is that you may both be happy. James must expect you to think of home and near friends you have left behind sometimes. If you could forget us so readily, I should fear you would not make a very affectionate companion.

The neighbors and friends are all inquiring about you and sending their best respects to you. Letitia says you agreed to write her, and Let was always one of your best friends. She would not do anything to hurt your feelings towards her. I know she always thought you and James both about right. She had nothing to do with that little affair. I know what made you think she did, but it was not so. Rose thought of nothing but having a little fun.[9] You say you get no letters. We have wrote as many from home as we have got.

[From Jenny to her parents.—Ed.]
Thompson's Station,
Tennessee, April 10, 1866

Dear father & mother,

I wish to know why you do not write oftener to me. I have had no letter for some time. The date of the last was the middle of March. Tell Lem & Lizzie that I wish them much joy, long life, health, and any amount of happiness. Tell me if you please, when they were married, if Lizzie had her children with her.[10] You know I would want to know all particulars in regard to such an event. I have been quite busy for the last two weeks. Let me tell you what I have done. Two weeks ago last Sunday, according to promise, we went to see Will and Fie, lost our road, and before reaching [the] house, traveled ten miles. When we got there, found Fie had not been out of her bed for a week, was feeling better when we came away. Will was obliged to go to Nashville & Fie wanted to come and stay with me. I was frightened at the thought, but concluded that it might do her more harm than good to stay at home when she wanted to go. We reached [the] house shortly after sundown, and the next morning made preparations for her coming. She came a little after noon, went right to bed. I had the teakettle boiling, made her a cup of tea strong enough to bear up an iron wedge. She drank three of them, and they rested her a good deal. We had a first rate visit. Will got home from Nashville Thursday. I did not think her able to go as she had not sat up a half hour the whole day. But he would not listen to either James or I. He thought she would be better off at home with him. We took a straw bed, laid it in the bottom of the box with a bolster at her back, and away they went for home. Will was here last Monday, said she rode home very easily. She is very nervous and I think ought to take something for nerves. I wish I had some nerve root to let her have. She would never have been in this way had she led a life of single blessedness. But I commenced to tell of my work. Mr. Critz came the night before Fie went away and was here [for] breakfast, supper, and nights until after noon Saturday. Sunday after getting dinner, my girl gave out and left me alone to see that the work was done. I got up Monday & went to washing, finished my washing before dinner, and after doing up my dishes went upstairs to work, swept the chambers & made the bed below anew as it was mussed. Dusted the things, looked out of the window to see if the coast was clear, when who should I spy but Jimmy, crossing toward the house with a couple of gentlemen, L. G. Kies & Philip Lander.[11] I retreated, leaving the coast clear. They stayed one night only, circumstances not thought of before hurrying them back Tuesday. After they were gone, I went to the garden with James, dropped the sweet corn and the popcorn Etty sent Jimmy, he says every kernel must go in the ground. Yesterday I cleaned some in the forenoon, planted fifty hills of cabbage, 7 of tomatoes, sewed 1 hour and a half, went to the field at half past 3 to see Jimmy marking cotton ground. Did not come up until he did, after sundown. Martha was well enough to help me yesterday and is to work today, though not hard.[12] We have had quite an increase in family since I wrote before. The old cow has a calf, and we have a squirrel and rabbit. They are running around the room while I write. We call the squirrel Bunny, he sleeps with Mr. Squire. I cannot write more now. Jimmy will write in a few days. Your daughter, Jenny.

[From James to his stepmother, Eliza King.—Ed.]

Thompson's Station,
Tennessee, May 13, 1866

Dear mother,

We are all enjoying as good health as usual and "live quite contented and free from all strife." Should I tell you mother that not a cross word or look has passed between us since our marriage, despite the many warnings I had

Thompsons Station Tenn
April 10 1866

Dear Father & Mother
I wish to know
why you do not write oftener to
me I have had no letter for some
time The date of the last was the middle
of March. Tell Ben & Lizzie that I
wish them much joy long life Health
and any amount of Happiness. Tell me
if you please when they were married if
Lizzie has her Children with her You know
I would want to know all particulars
in regard to such an event. I have
been quite busy for the last two weeks
let me tell you what I have done two
weeks ago last Sunday according to promise
we went to see Will and Sis lost our road
and before reaching home traveled ten
miles when we got there found Sis
had not been out of her bed for
a week was feeling better when we came
away Will was obliged to go to Nashville &
Sis wanted to come and stay with
me I was frightened at the thought but
concluded that it might do her more harm
than good to stay at home when She wanted
to go we reached home shortly after Sundown
and the next morning made preparations

Jenny's letter to her parents, April 10, 1866. (Courtesy Western Michigan University Archives
and Regional History Collections)

to the contrary previous to that event, you could hardly believe me. Even you told me "that I did not know Jenny as you did," and one of my friends who has passed the age of forty told me that we would quarrel like cats and dogs after six weeks had passed away. The quarrels foretold have not come, and I must say that I am tonight a happy man with a happy, loving wife at my side.

I see by the tone of your letters that you wish me to come back to Michigan and settle down. This I can never do, so long as matters remain as they now are. You and father has done many kindnesses for Jenny and I, but there are others who ought to be near and dear to me who have wronged me, and until they acknowledge that such is the case, I shall never trouble them more. Not that I cherish any hatred toward, or would do anything that would injure them, but "there is a time when forbearance ceases to be a virtue," and rather than ask a favor or help from them I will saw wood the rest of my days. But on the contrary, should they ever come to want or need assistance, I would lend a helping hand were it in my power. This is the reason I do not want to visit Michigan, and the reason why I never shall make it my home. If Jenny wishes to make you a visit and there is an opportunity of her having company on the journey, I shall urge her to go, but it will be a long time, if affairs remain as at present, before I do.[13]

How I wish you and father could make us a visit this summer. I am certain you would enjoy it. Should like to see [you] all very much. Tell Etta I have hoed the popcorn once, and it looks finely.

Farming moves along as well as I can wish. Most of our cotton is up, and we shall go to working it tomorrow. One piece of our corn we have worked one way, and half through the other. Jenny, I suppose, will tell you all about the garden, cow, affairs about the house, &c, &c, so that I will bring mine to a close. Hoping to hear from you soon, I wish you a kind goodnight. James.

———

Jenny returned to Michigan in early September, midway through her first pregnancy, seeking the comfort and security of her extended family in Fabius. Her separation from James triggered a flurry of letters. The contents offer insight into how things were progressing on the plantation, as well as how the Kings viewed the political situation spiraling out of control in Washington, where President Andrew Johnson was clashing with radical Republicans in Congress over the course of Reconstruction in the South.

———

> Home without Jenny,
> Thompson's Station, Tennessee,
> September 13, 1866

My dear little wife,

After leaving you at Nashville, I went to Maj. Kirk's office and stayed all night with the boys. Webb [Daniel Webster Holbrook] was already dressed to come and see us, thinking that we had stopped at the hotel. He was sadly disappointed. The next day I returned home, and dear Jenny, need I say that I was lonesome? Kies left us today to go to McMinnville, will be back sometime next week. Thomas Iddings is here, came day before yesterday.[14] Linus is at the barn, working at the press. I am upstairs. Oh Jenny how much I miss you. I opened the press door yesterday and your sundown was the first thing I saw, and if it had been a thing which is possible for me to do, [I] should have had a good cry. Sometimes I come into the house and find myself halfway upstairs going to see Jenny.

The boys picked 218 lbs. of cotton yesterday.

If you have not yet written me, do so by all means, your first opportunity. Your affectionate James.

> My lonesome home,
> Thompson's Station, Tennessee,
> September 15, 1866

My dear little wife,

It is 8 o'clock, our usual bedtime, and I am here alone to occupy the bed where you and I so oft have slept. Jenny I don't think any other man ever loved his wife as I. I am up in our room, seated by the stand. When I came up, I took up the Bible to read a chapter, and I found inside its pages your hair and mine that you so fancifully arranged. Never again, dear Jenny, will I consent to a separation, let the consequences be what they may. Without you, life is a blank.

Your letter from White Pigeon I received this evening, and of course relieved me of much suspense, and tells me that soon you would be among friends. I almost envy them the joy of meeting you, and Jenny after I bade you adieu on the train, I had thoughts of asking you to stay with me. Don't share this to mother or anyone. How selfish she would think me.

Sunday 16th

Last night, dear Jenny, I dreamed you were with me, but awoke this morning to find a long Sabbath to be passed alone. We have just been to breakfast. Harriet is doing the work.[15] Linus is dressing himself. He says to me, "Boy, this will be the lonesomest day that you have passed for a long time." Tommy Iddings stayed with us until yesterday morning. Linus says that "we will stay at home until we get so lonesome we can't stand it, then go over to Billy's or ride somewhere." We have 1,000 pounds of cotton picked. This next week we will have to work on the road, therefore will not get much picked. Kate and Crockett had another flare up yesterday, but this morning are alright, I think.[16] We have the press nearly finished. I wrote Mr. Baur in regard to my 40, telling him that he could have it by paying $700, and 300 the coming winter.[17]

Jimmy has just drove the cow into the yard for Harriet to milk. Lawson and Julia are running around as usual.[18] Linus received two papers from home last night. One, the New York Tribune, and the other, the Detroit. Nothing in them but politics and Negro suffrage. Let them come here & live among them, work them, and be run over by them, and their everlasting cry would be stopped. When anyone talks Negro suffrage to you, tell them your experience and don't be afraid to talk.[19]

It is three o'clock, and Harriet is getting supper. Have been asleep nearly all day. The day you went away I weighed 122, today I weigh 116. Can you tell me now what's the matter? You know what you used to tell me. Don't tell our folks what I weigh, or they will be sending you back right away. Before I come home I will try and gain some, or they will think I am a small specimen of a man. It don't matter much, as long as I feel as large as anyone.

The boys saddled the pony and our mule. Linus and I rode over to Will's. Found them all well. Fie is real smart. They make considerable fun of me being left a widder man, but it ain't no laughing matter.[20] Are you as happy at home as you used to be with me Jenny, tell me truly. Fie said to me tonight, "Do you suppose Jenny will ever be contented to come back and live with you here?" I told her you would be content with me in Africa or any other place.

Monday, September 17

The boys have gone to work on the road. Linus is at work on the press. Harriet keeps things pretty snug so far. Don't know how long it will hold out.[21] Write as often as you can. Tell mother to write me, & Dio [Darius].[22] James.

Thompson's Station, Tennessee,
September 17, 1866

My dear little wife,

This morning I carried a letter to the office and mailed it to you, and now I have commenced another. A part of my errand to the station this morning was to get some large nails to use about the press. I could not get any there, so came home and rode over to Mr. Shields to see if he had any.[23] Was disappointed there, returned home and made wooden pins which answered every purpose. The boys worked on the road today. Mr. Shaw has been here and took supper with us since you went away.[24] I forgot to tell you that the day you passed through Nash[ville], the cholera was raging. A man died at the St. Cloud Hotel while I was there.[25] There was some 50 cases in the city yesterday. Webb and the other boys take their regular cholera medicine. I slept with Webb the night I stayed there. He is not half as good a bedfellow as you. I told him so. He was really disappointed at not seeing you.

Tuesday, September 18

Today Linus and I worked at the press. Went to the cars at mail time, thinking I would get a letter from my dear Jenny, but there was no mail on account of the agents taking the cholera. There was 40 deaths in Nashville yesterday from this dreadful disease. I fear for Webb and the other boys.

Mr. Critz came to see us and is here tonight, he has just gone to bed.

Jenny I have made up my mind, if I have any capital to work on when I come out of this, to buy me a little home, settle down with my Jenny and take comfort, for she's all the comfort I care for in this world.

[A fragmentary letter follows. The first six pages are missing.—Ed.]

[Thompson's Station, Tennessee,
September 18, 1866]

[I] have turned the bed down and have been looking for the revolver, but cannot find it. Harriet is the worst girl about putting things away I ever saw. Write often Jenny, and great long letters. Tell me everything. Whether J. Webster behaves himself as he should &c.[26]

Wednesday, September 19

This is a rainy morning. The boys got ready to work on the road, but were driven off by the rain. Mr. Critz remained with us until after breakfast, and then went over to Tommy's. He is getting scared about the manner the president was received throughout the North, and wants to sell his place. He says he believes that the day will come when every foot of the land owned by men who participated in the rebellion will be confiscated.[27] I told him that he was more scared than hurt. I believe that if we have a true friend in the South, he is the one. Linus is going to the sawmill. Harriet is doing the morning's work. What is my little Jenny doing this morning? Have you thought of poor me all alone? Have written two letters this morning. One to [Ephraim] Hall, the other to Mark Price.[28]

"Joy to the world," I've found the album. Harriet put it in your bandbox for safekeeping. I took out the album, took a long peep at my Jenny, then took up the box to put it up, when I saw an envelope containing something. I opened it and found two certificates of marriage.

"Happy they; the happiest of their kind!
Whom gentle stars unite; and in one fate
Their hearts, their fortune, and their being blend"
"How blest the sacred tie that binds
In Union sweet accord and minds
How swift the heavenly course they run,
Whose hearts, whose faith, whose hopes are one"[29]

Josh has just returned from the station and brought me two letters, one from Dio, and yours written while at the crossing.[30] The one from Dio is dated the 13th, tells me of your safe arrival, and says that you like it much better in Michigan than here. Does my little Jenny like it so much better there than here that she would consent to leave poor me here alone, and she stay there? What would you do for kisses, and whose arms encircle you in your slumbers? I will bet Jenny, if the truth was known, that you have wished yourself back with me a dozen times ere this. Here is one little thing in your letter which settles all Dio says. "Jimmy, more than half the time when I drop asleep I am waked, how? I find my lips reaching for a kiss. Oh how much I miss those arms no one can know."

Evening has come once more. Squire Critz has come back. I have been singing for him. Linus played on the fiddle. Goodnight dear Jenny, may the good angels guard and protect you.

[From James to his older brother Henry.—Ed.]
Thompson's Station, Tennessee,
September 19, 1866
If you can't read this, take it to Jenny.

Dear brother,

Yours of the 13th inst. came to hand today, and I am going to do better than you, be more prompt in answering. Since Jenny went away, I have hardly known how to content myself, and the hours seem days, the days, weeks. Our cotton is opening rapidly and is very good picking, although we are doing nothing in that line this week. All our hands were ordered out on the road by the overseer of highways, and things are at a standstill. The overseer wishes me to work. I told him I was not of much account nohow, and would send a mule team and plow in my place. To this proposition he readily consented, and I and Squire are at work at our cotton press. There is nothing in the way of news. I see by the Northern papers that politics are raging to a fearful extent. What do the people at home go for as a general thing, are they radical and down on Andy Johnson? I am no Johnson man, he is too big a dose for so small a man as I to swallow, and radicalism in every sense is hard to get down. Any man who goes for extending the ballot box to the sons of Africa at this present time is woefully in the dark. Let any person whose wishes are such come south and have the management of them, and see the animal in his native lair, and if their philanthropic ideas of suffrage don't collapse, then I'm no prophet.

The cholera is raging dreadfully in Nashville. There were 50 deaths from it yesterday, and the number of cases [is] increasing. Many of the citizens have left, business houses have closed, and dullness reigns in Nashville for a time. Out here on the plantation, we have no fears. It cannot live in the pure air of the country.

Then Jenny likes it better at home than she does here. Am sorry to hear that—am fearful if such be the case that I am elected for another term of "Bachelor's Hall" that I had not counted on. I wish you would take particular notice and see if she don't sigh once in a while to come back to Tennessee. I expect to make this land of the South my home for some time to come, and hope that I shall not have to live long at a time without Jenny, because she is a little body that can cheer a fellow up and put the bright side of the picture out when a fellow gets the blues.

We have 1,000 pounds of cotton picked, and next week will go at it in earnest. How would you like to come to this sunny land and help me run a plantation? You ought to see the country at any rate, and if Jenny and I live in the South next year, you must stay with us, what think you? James.

Thompson's Station, Tennessee,
September 20, 1866

My dear precious little wife,

It has been a rainy, dreary day. I went to the station this forenoon and carried a good long letter to the office, expecting to get one from you in return, but must confess Jenny that I was sadly disappointed. Why did you not write me Jenny the night after you got home? I think I can see the whole thing. Weren't you sick Jenny, unable to write, and is not that the reason I receive no letters

from you? If you are sick dear Jenny, do not try to disguise the fact from me. Have mother write and tell me all if you cannot.
Saturday afternoon

Will's folks are very well. Tommy started for home today. I expect you will hear of his arrival in Michigan ere this reaches you. Linus has got the dumps this afternoon and don't say much. I am lonesome.
Sunday eve

Last night about dark, a covered carriage drove along the road and stopped at our gate. The gentleman hallooed, I went to the gate, he inquired if Mr. Squire & King lived here, I told him they did, and invited them to alight. They did, and it proved to be Mr. Recksford and lady. He formerly practiced law in Detroit, Michigan, but is now practicing at Nashville.[31] They are here now, will probably remain until tomorrow. They are very pleasant people. The colored people all say that she looks exactly like Miss Jenny, but I don't think her half so good. Harriet got a nice dinner today, and keeps things about the house as clean as can be. Since you went away she has scrubbed 3 or 4 times upstairs and below, mopped the hall twice, and cleaned up generally.

The darkies are having a prayer meeting, and two or three of them have got the power. They make fine music. Mr. & Mrs. Recksford think that is too much of a good thing.[32]

Night before last, we had a slight frost, which done no injury. Am in hopes that we shall have no more until our cotton is out of the way. We are going to picking in the morning in earnest. Our press is nearly completed. Jenny until I received some word from you, is there any use of writing more? Your James. It is Sunday evening, Mr. & Mrs. Recksford are reading some of our books. Linus sits with his chair canted back against the wall downstairs, saying nothing. Aunty and the colored population in general are all well.[33]
Harriet has forgotten to make my bed today and I am forcibly reminded of,

> Late when he goes to bed at night shivering
> He crapes like a terrapin under the kivering

How long do you think it will be before I can come for you, or couldn't you positively state?

Never a bit is the bed made at all[34]

Don't let anyone see this.

Thompson's Station, Tennessee,
September 23, 1866

My dear little wife,

Yours of the 16th came to hand today, having been on the road 7 days. Now, my little Jenny, I had become somewhat anxious on your account, as you have perceived by my letters.

Now Jenny as regards the danger of living here, there is just none at all, and I shall come after you this winter, surely. If I had the means to buy father's farm without running in debt 3 or 4,000 dollars for it, I would be satisfied, but you know that I have not. I wish father could come to Tennessee sometime this fall and look at the country with me, and then if he advised me to live [in the] North I could be persuaded to do so, but Jenny in this climate with a farm as large as father's, I could clear double the amount from it that you can in Mich. If he will come here sometime this fall and look around with me, I will pay the expenses of the trip.

The boys picked 690 pounds of cotton today, we now have 1,690 pounds of seed cotton.

You need not think that I will stay away any longer than my business requires. Do not urge father to come here if his business will in any way suffer by his absence, but I do wish he was here and I could soon decide either the one way or the other.

Linus has got the mully grubs tonight, I guess.[35] He don't say much, what confounded spells.

How can you like such a self-willed, headstrong fellow as my folks say I am, and one who wants you to live with him in Tennessee or some other part of the South?

All that I feel hard toward him [father] for is this. In the first place, he urged me to take the 40 and I did not want it.[36] He then made as fair an agreement as was ever made in the world, to let me have the 40 for 500, and 500 to come out of my portion of the estate if ever divided. I sent my money home to him after coming south, and he used it up without my knowledge and then coolly informed me that he had used it and wanted me to turn it on the lend. I am going to write him tonight and tell him all about it, and that he need not blame you. Had he have let me known that he stood in need of the money, every cent would have been at his disposal that I could command, and after using my money, he shamefully abused me at the breakfast table in the presence of Uncle Anson and the others. I can't write any more tonight. Your ever loving James.

Tuesday, September 24th

My little wife another long day has passed away. Will Iddings came this forenoon and stayed until after dinner, then went home. Fie is feeling pretty well. She says that you never will leave your folks to come south again and live with poor me. She don't know you as well as I, does she Jenny? We have just been to supper. Linus is reading some old history that he picked up. Now as regards the selling of my 40. I believe if I had 6 or 700 to work on now, that is, buy cotton in the seed and gin ourselves, I could clear a right smart; and I have given my word to Mr. Baur and cannot conscientiously withdraw it until I have given him ample time to answer. It is now nearly 3 weeks since I wrote him.

Our boys have picked 320 pounds of cotton today. It commenced raining about 10 o'clock, and we had to hole up. Robert picked 150 pounds yesterday in about 8 hours.[37]

Thompson's Station, Tennessee,
September 25, 1866

My dear little wife,

I have been hard at work filing the gin saws. I filed 19 of them today. My health is pretty good, though I think if I could see you, it would improve. I told you that I would write father, but Jenny I have made up my mind not to, and I want you to see him and talk plainly to him. The boys picked 767 pounds of cotton today. Squire Critz came and stayed a little while with us, and then went to Tommy's.[38] I received a letter from Uncle Anson, and he strongly insists on my not going to Arkansas. Edwin Dickinson says, "Tell him to keep out of that God-forsaken country."[39] I can tell between this and the time we break up here where I will settle, and Jenny, wherever that is, I want you with me. Jenny if you think I am a supporter of the president, you are mistaken. I have not been for some time. Neither am I radical. You know my views on the question of suffrage. I do not apprehend any trouble in this section. If I did, I would not think of having you come back. I can't write more now, goodnight dear Jenny.

Thompson's Station, Tennessee,
September 27, 1866

My dear wife,

In the afternoon Squire Critz came here and stayed all night with us. He went away this morning. I have been at work filing gin saws, will get through tomorrow. The boys picked 706 pounds of cotton today. We now have nearly

4,000 pounds out. After dark, Linus got out the violin and played a few of his favorites, while I lent a vocal accompaniment. Among the pieces was "Maggie Dear." Wish Dio had been here to help me. I hear Linus in the other room, just going to bed. Poor boy, the women never trouble him, so he says, but then I doubt it the biggest. I received a letter from Webb yesterday. The boys are all well. Maj. Kirk has been sick over two weeks.

Sunday, September 30

Jenny do not think that I support the president in every sense of the word, for I do not. When he vetoed the Civil Rights Bill I think he done a good thing, what any sensible man would have done, but when he makes an ass of himself as he has been doing for the past 3 months he is not doing so good a thing. Some of the people here look for his impeachment as soon as Congress assembles, and are looking for the most rigid laws to be enforced.[40] Mr. Critz thinks that the day will come when all the land in the South will be confiscated. Now Jenny such men as he I sympathize with, and if I can ever lend him a helping hand, [I] will do it. I do not think that he holds any prejudice against Northern men, and would aid me as quick as any Tom. Critz came over yesterday to see if we would gin his cotton for him. I think we shall clear some money off our gin.[41]

Mr. Kinkade has just been here to get the loan of a mule to ride over west.[42] Linus told him he reckoned he could have one.

Some of the boys are going to church. Some time ago you recollect that Crockett made a bedstead, and he and Kate had quite a rumpus over it. Yesterday a woman came here and told Kate that Crockett made it for her, and that she was going to stay all night. To this Kate would not consent and more than went for her, and I believe finally drove her off. Crockett had nothing to say in the affair. Let them fight it out on their own hook. They are a curious nation.

I see by the Detroit paper you sent that Col. Stoughton is nominated for attorney genl. of the state.[43] Wonder if he has forgotten the boy who watched over him night and day?

The darkies have had their regular meeting today. There were a number of females among them. Crockett done some tall praying and I think went home with some of the wenches after services. Have you told mother their peculiar styles down in this country?

I am just the lonesomest little chap tonight you ever saw. Then mother thinks that J. Webster will want to see his father, I don't doubt it in the least. This is Monday morning Jenny, and I must go to work. Will start the gin today.

William Lewis Stoughton. (Library of Congress)

Thompson's Station, Tennessee,
October 3, 1866

Dear darling little wife,

The boys picked 694 pounds of cotton in the upper field today. Manda, Melvina, and Kate are helping. Just think of a woman as far along as she stooping over all day, picking cotton.[44] We have upwards of 6,000 pounds picked tonight. We finished the press today and will try it tomorrow. Linus says that Jim told him Mrs. Kinkade has got her leg broke again.

Thursday Eve, October 4

Well today we have ginned some, and tried our press. We did not get a bale pressed on account of one of the ropes breaking on the pulley block. Because we did not succeed, Linus has got the mully grubs again. Today the boys picked 794 pounds in the upper field. Tonight, Jenny, we have 7,044 pounds picked.

Friday morning. We are going to fix our press to work with levers. I do not think, Jenny, there is the least doubt about our making a little something in the cotton crop. Have patience dear Jenny, I will be with you as early as I can. Your James.

Thompson's Station, Tennessee,
October 5, 1866

My darling little wife,

Your long letter dated Sept. 30th, and mother's enclosed, reached me today. I think you will be confined in December myself, but there is no telling, you may not go longer than November. At any rate, I shall try and reach Michigan before that time. The boys picked 840 pounds of cotton today. Manda picks all the time. It would kill anybody else beside a darkey.

Jenny I am careful. Every night I sleep with the revolver under my head. I do not think there is more danger here than in Michigan. The people all continue to treat us friendly, and as long as they do that, I shall do the same by them.[45]

Sat. morning before breakfast. We are going to run the gin today, have about 4 bales ginned. Linus will get the press fixed so that we can finish pressing the one we commenced day before yesterday. Can't write any more this morning. Your James.

———

On October 6, the Kings' first wedding anniversary, ex-Confederate lieutenant general Richard Stoddert Ewell visited James and Linus's plantation. Ewell had suffered a cotton press failure, and wished to observe how the Northerners on the Critz farm were handling their press. Ewell and his wife had settled at the end of 1865 on her three thousand-acre farm—one of the largest in Tennessee—at Spring Hill, just south of Thompson's Station. Ewell did not believe in holding grudges over the war, and exhibited little patience for those who did. Lieutenant General Richard Taylor said of Ewell, "Virginia never bred a truer gentleman, a braver soldier, nor an odder, more lovable fellow."[46]

———

Thompson's Station, Tennessee,
October 6, 1866

My dear little wife,

How many times today have you thought of one year ago.

Happy they! The happiest of their kind!
When gentle stars smile; and in one fate
Their hearts, their fortune, and their being blend

Yes dear Jenny, one year ago today saw our happiness complete and never for a single moment have either of us had cause to regret the fate that joined us. Tonight finds me in my room as usual. Linus has got out his violin and [is] trying to drive dull care away. May the next 6th day of October find us enjoying each other's society and not as this, so widely separated.

Jenny I am going in bed and think about one year ago, are you doing the same?

Sunday Morning, October 7

Harriet walked 5 miles to church last night and back again, making 10 miles in all, got back this morning at 4 o'clock. She has got her work all done, excepting milking the cow. She generally puts this off until the last day in the week. We have got over 8,000 pounds of cotton picked. They have picked about 5,000 out of the upper field, where some of the people thought the Yanks on Critz's farm would not make their rent.

General Ewell came here yesterday to see our press. He is a pleasant-looking man, and does not seem the fighting character he really is. He lost a leg in the service and is now stomping it through the world. There was one General Harlan with him, someone I had never heard of before.[47] They were anxious to see our press work. Said he would come again.

My dear Jenny, Linus and I started out for a ride at 12 o'clock, he mounted on the pony, and myself on the mule. We took the station road and went as far

Richard Stoddert Ewell.
(Guernsey and Alden, *Harper's
Pictorial History of the Great
Rebellion*)

as the turn there. Turned and passed Mr. Trimble's, Ridley's, Dr. Steel's steam
sawmill, and went over on to the Carter's Creek Pike, followed the road until
we come out by Mack Drake's, then went around by Will's.[48] Found them all
well. Fie says she and Will get along first rate. Stayed there to supper. When
we got home, Harriet had supper awaiting us, and of course we had to eat.
Then Linus played on his violin, and I sang. Has Dio got Tennessee fever any?
Goodnight my loving wife. James.

Don't care much about selling [the farmland in Fabius], Jenny. I believe we
shall make something here, and in six months or a year, I can buy land here
cheaper than at the present.

Thompson's Station, Tennessee,
October 20, 1866

My dear little wife,

Have been ginning today. I run through 500 pounds of seed cotton in 52
minutes. We can gin 4 bales a day very easily. This is very fast work. They
generally run through about 2½ or 3 bales per day. Our press is going to work
admirable.

Sunday Evening October 21

This morning Linus and I saddled our beasts and went to church at Pope's Chapel. Heard a good sermon preached by an old gentleman named North.[49] After services, some eight or ten were baptized. When church ended, we rode over to Will's. Did not get started from Will's until after dark. Soon after I left them, the rain commenced pouring down, and before I got home [I] was pretty thoroughly drenched, came through by Col. Helms.[50] Aunty had a good fire in the fireplace, and I do not feel any the worse for my soaking.

If I can possibly start for home the middle of next month, I shall do so. Am feeling pretty well this morning. Shall run the gin today. Write as often as you can, dear. Your James.

Thompson's Station, Tennessee,
October 22, 1866

My dear beloved wife,

Oh how anxiously I am looking forward to the time when you will be restored to my arms, never to be torn from me again in this world, I hope. Were only my own means concerned in our business, I would leave here tomorrow, but I do not feel that I could consistently take such a step. Feeling must follow in the wake of duty. I think by the last of November I shall be in Mich. Oh dear Jenny, do not worry on my account. Only think how much more reason I have to feel worried. The condition of your health at the present time. May God help you dear one. Jenny what would become of me if you were lost to me? I would not care to live. Is this wicked, I do not believe it is. You are my all, in you are all my hopes centered, and life would be an aching void Jenny. A truer heart than mine never beat in the breast of man. For you I lived, and for no other, and when I gave it to you, I knew I was getting as true in return.

Tonight we have 15,550 pounds of cotton picked.

We got up this morning expecting to press 4 or 5 bales and had so much bad luck that we only got 1 pressed. I am tired. Write me just what Vet says about my 40.[51] You may look for me in Mich. by the 27th of next month.

Wednesday 24th

Last night we had some frost. All the cotton will be opened in a week. Shall run the gin today. Give my love to all. Your affectionate James.

Thompson's Station, Tennessee,
October 26, 1866

My dear little wife,

Over one week has passed away since receiving a word from you. Now you must know I am getting somewhat anxious. Last night I went home with Will and stayed all night. Billy's hands all left him Monday morning. They did not earn their board, and Billy told them they could either pick for $1.00 pr. hundred or they might work for $10.00 pr. month until Christmas. They would do neither and demanded their pay, and declared their intentions of leaving. He told them that if they went they would not get a cent of pay, but if they would stay with him and work as they agreed, every cent coming should be promptly paid. They went and made complaint before a justice of the peace at Spring Hill, and this morning, Mr. Constable served a summons on William. The trial comes off tomorrow. I think they can do nothing with him, as they did not fulfill their part of the contract.[52]

I took 2 bales of cotton to the station yesterday and intended to sell it, but they only offered 29 cents, and I thought I would wait a few days and see if the price does not increase. Shall work at baling tomorrow. Today we have been ginning. Tonight we have 17,340 [pounds] of cotton picked. We bought 500 pounds of Mrs. Kinkade today, and paid her 8 cents pr. pound.[53] Nancy is very anxious that you should come back.

Sunday Eve, October 28, 1866

My dear Jenny this is Sunday night, and I am seated by the fire writing. Jim Porter came from Nashville to make us a visit, and today we went to Will's.[54] He returns tomorrow morning.

Oh what shall I tell you, dear one? I have not received a letter from you in 12 days, what is the reason? If you are sick, write me and I will come to you. Do tell me all, Jenny. James.

The suit was decided in Will's favor.[55]

Nashville, Tennessee,
October 30, 1866

Dear wife,

This night finds me at the Stacy House in the city. I came in on the 7 o'clock train. Will Iddings came with me. He has gone to the theatre. I did not choose to go. Shall go out and see Webb in the morning. I came on purpose to pay Price. I sold 2 bales of cotton today for 29 cents pr. pound. Got $260.42 for the 2.

Jenny I often make up my mind to come to you, let the consequences be what they may. If I could only hear from Kies, I would leave right away. Am coming home to sell my land, and buy me a home where I can get land cheap, either west Tennessee or north Ala.

Land can be bought very cheap now, and I think if I ever get me a home, now is the time to make a strike. Oh won't we take comfort when we get settled down again, you and I and the little one?

The 16th Regular band have just been playing in front of the hotel. Shall go back to the farm tomorrow. Goodnight, my own dear wife. James.

Thompson's Station, Tennessee,
October 31, 1866

My dear little wife,

I have just returned from Nashville, where I went to pay Price. When I got there, he told me he did not want to use the money, and to keep it as long as I wanted to. So I kept it. Am going to buy cotton with it.

Nov. 1st.

Have been ginning this forenoon. Mr. Critz took dinner with us today. We have 20,000 pounds of cotton picked, and I think about 6 or 7 thousand in the fields. The middle and corn fields were injured to a great extent by the rust.

Tell father if anyone wants my 40 that he can tell them they can have it for $1,300. Your James.

[From James to his stepmother, Eliza King.—Ed.]

Thompson's Station, Tennessee,
November 2, 1866

My dear mother,

Yours & Jenny's letter of the 28th of last month came to hand today, and found me feeling very well considering everything, for you must know that with my dear Jenny so far away and knowing the delicate condition of her health at the present time, my mind is continually uneasy and worried. I know, mother, that you are kind and can do for her better than I could, perhaps, but still something tells me the great danger she is soon to pass through may not be safely passed. Were only my own interests concerned in the South, I would be with my Jenny as soon as the trains could carry me, but as it is, "feeling must follow in the wake of duty." I shall be in Three Rivers the 27th of this month, no bad luck. How thankful I am that my Jenny and I am blessed with so kind a father & mother, and someday all kindnesses shall be repaid.

Now mother, as regards our cotton experiment. I do not think we shall make any money. At least, the chances are against us. Two of our fields were badly injured by rust. If I come out with money enough to pay what little I owe, I shall be thankful. Am going to close the thing up as soon as I can come home, sell my land, and as soon as Jenny is able, buy me a home in west Tennessee. Good improved lands can be bought for ten dollars an acre. Now do not say anything about my affairs to any but father and Jenny. I do not want Mich. people to know anything about them. It does not concern them at all. Am not sorry that I tried raising cotton this year. My experience is worth a year's hard work.

One of the hands (Harold) has just come in and wants me to write him a love letter, and I expect I shall have to do it. He has left her name and says "Mr. King, fix it up mighty nice. I know you can do it. Write her just such a letter as you used to write Miss Jenny."

Hoping that not many weeks will pass away before I can greet you all at home, I remain your affectionate son, James.

Thompson's Station, Tennessee,
November 4, 1866

My dear wife,

This has been a lonesome day, although Linus and I went to Will's and stayed a good part of it. Fie got supper for us. I told her you did not like it because she wrote that you were homesick, and she and Will both said that you were very, and that she had not been a mite. I told them they wanted to tell that to marines. They could not get that down an old salt. And I knew that if F—told the truth, she would say she was. This rather took them down, and they did not want to talk about homesick people anymore. I know too much of their affairs to make me believe anything of the kind.

I am going to Nash[ville] tomorrow to see one of the boys about going to west Tennessee with me, and if I succeed, shall go right away, and then I am coming home. Don't you wish you could read shorthand?[56] Then I could write just as big a love letter to my little Jenny and no one could break open and read, wouldn't that be nice? Shall we teach the other what is to be how to write love letters, or ain't you going to let him love anyone? Won't he think lots of his nice little mamma when he gets to be a big J.W., I'll bet he will. Goodnight. Your loving James.

[From Linus Truman Squire.—Ed.]

Thompson's Station, Tennessee,
November 14, 1866

Dear James,

Enclosed please find letter rec'd. yesterday. All well. Business dull. Raining a little today. No cotton to gin. Cotton dull yesterday in Nash[ville] at 28 1/2c, 29c tax paid. Haven't sold any yet, and can't buy until I do sell some. Shall start out soon and see if I can get any to gin. Prospect dull.

Iddings's people [rode?] Sunday.

Today is my birthday, so please excuse me for not writing much, as I want to celebrate it by putting the chains on the head block and pressing a bale. Linus.

———

Crop failure rendered the entrepreneurship at Thompson's Station unprofitable and postponed James's plans to buy a home and settle down, but there was cause for rejoicing when James and Sarah Jane King's first child, Jennie May King, was born on January 2, 1867. Later that year, the Kings brought their infant daughter south and made another go at cotton farming, this time about five miles south of Decatur, Alabama.[57] Decatur was all but destroyed during the war and suffered its own civil war in miniature as local Unionists and Confederate sympathizers resorted to guerrilla warfare and inflicted untold depredations against their neighbors. Murders, and general lawlessness, remained frightfully common for some time even after the war ended. Few local newspapers and government records from this period are extant, and with the exception of a single letter, none of the King correspondence from 1867 survives.[58] James and Jenny apparently spent the entire year together, and that goes a long way toward explaining the lack of letters for this period. But a tantalizing glimpse of their experiences in Alabama that year can be reconstructed from other sources.

The Kings again placed all their eggs in one basket, and they could hardly have chosen a worse time to run a cotton plantation. In 1867 planters suffered through the worst crop season in memory, spelling financial ruin for many. To top it off, the political climate turned even uglier, as the Republican-dominated Congress, enjoying the mandate granted by the landslide 1866 elections, came down harder on the South, with the Military Reconstruction Acts in particular. Many Northerners who had arrived in the postwar South to civil greetings now found their presence completely unwelcome and feared violence. Most carpetbagger plantation lessees streamed back to their homes in the North and never looked back.[59]

Jenny and May returned to Michigan ahead of James. The Kings' only surviving correspondence from Alabama is a brief letter James wrote to Jenny shortly before he left the South for good. There are, however, scattered clues regarding the exact nature of the family's ill fortunes. James declared that he "returned to Michigan twenty dollars worse off than nothing, in a financial point of view." For the rest of his life, he compared his second season of cotton farming with the experiences of Albion Tourgée's protagonist Comfort Servosse in *A Fool's Errand*, a classic 1879 novel based on the experiences of its author in the South during Reconstruction.[60] James later wrote that many of his friends credited him

> with the authorship of that work. Its real author Judge Tourgée was unknown for some time after the publication of the book, and the late Judge M. V. Montgomery of Lansing said to [me], soon after the work appeared, "James you need not deny the authorship of the 'Fool's Errand,' for there are more than a dozen incidents in the book that you related to me long before it was published." The reply was, "I would like the honor of writing such a book, for it is a masterpiece, and as a participant of Northern life at the South, at the close of the war, it is true to living reality. Whosoever the author may be, he knew what he was writing about, and in no way is the picture overdrawn."[61]

What can be inferred of James's own experiences from the events depicted in *A Fool's Errand*? Tourgée's volume illustrates the violent postwar assertion of Southern culture and politics in defiance of Reconstruction. The fictional Servosse was also a Michigander, but he was a lawyer and a radical Republican in contrast to James, a farmer and self-declared moderate. Nevertheless, many of the events taking place in *A Fool's Errand* may mirror James's stint in Alabama. Upon arrival at his plantation, Servosse was greeted with civility by his Southern neighbors—until he associated himself with Northern girls serving as schoolteachers for the freedmen. When a resident warned the transplanted Yankee to steer clear of the controversial educators, Servosse coldly rebuffed him. In retaliation, the carpetbagger was harshly denounced in the local newspaper as a "fanatical abolitionist" with "infamous doctrines." Thereafter, Servosse was urged by the locals to clarify his opinions at a large political gathering, and after indulging his audience with a most unwelcome speech in favor of freedmen's rights, he was waylaid in the dark on his return trip home. The protagonist found himself increasingly ensnared in local politics, vilified for advocating freedmen's rights and for selling them land, and was ultimately compelled to arm his family and their ex-slave associates in response to death threats. The Klan then arose with its reign of terror, inflicting torture and murder upon freedman, carpetbagger, and scalawag alike.

Decades later, a friend wrote in memory of James that he "went into the south to raise cotton. The Ku Klux clan drove him out with the loss of all his property." The crushing plantation failures of 1867 alone could have triggered the Kings' departure from Alabama; James later commented on having over-extended himself financially there. But the Ku Klux Klan was culpable for his abandonment of the South.[62] Stories of James's confrontation with the Klan have been passed down through generations of his descendants. Family tradition says that the Kings received repeated KKK threats, and that the tongues were cut from their horses' mouths one night. Another tradition holds that Klansmen once approached the Kings' plantation house in the middle of the night. James was alert and overheard his would-be assailants undergo a sudden change of heart—out of concern that their intended victim might have been awaiting them indoors with a revolver. The attack was aborted, but such imminent danger forced the family to flee with little more than the shirts on their backs.

Jenny and May departed first, while James stayed behind to tie up loose ends. On his last night in Decatur, he took refuge with two carpetbaggers who were running another local plantation: former captains William H. Williams and Ethan O. Hurd of the 39th Ohio Infantry. Hurd later described his and Williams's experiences with local Klansmen:

> One night on the plantation [Williams] awakened me under the impression that he had heard a Ku-Klux whistle, and we sat by the windows with our guns in our hands awaiting an attack that never came.
>
> There were Ku-Klux in the country, though. One night, when on the streets of Decatur in company with a Southern man, we met them in their masks and uniform, with revolvers drawn. They halted us, and demanded where we were going? We replied: "To church." They said that was a good place to go to, and they would go along, and escorted us to the Episcopal Church, where, it being just before Christmas, the young people were at work getting up evergreen decorations. They followed us in, and took seats in the rear. Some of the young ladies played and sang for them, and they then rose and left.[63]

At Capt. Williams's,
Decatur, Alabama,
December 29, 1867

My dear wife,

Nearly one long week has passed away since you left me, and there is no use of my telling you how anxious I am to hear of your safe arrival at home. Tonight, Jenny, I packed my satchel and left our home at the foot of the mountain, and tomorrow I take the train for Cincinnati.

Have not succeeded in selling pony & the cow. You know my circumstances and my needs, and if the friends at home can aid me any, perhaps I need it at the present time worse than I ever shall again. I borrowed $35 of George Holbrook to send you home.[64] I want to repay that as soon as I can. Oh Jenny how lonesome without you and May, and as soon as I possibly can, I must have you with me. Can't write you a long letter. Will write you again at Nashville. Have left Pony and the cow for Will to sell. Direct to Nashville, and George Holbrook will forward to me. Your James.

Do not tell my folks what I am doing.

———————

On January 8, 1868, James, returning from the South, wrote to Jenny with startlingly uncharacteristic bitterness:

You do not know the anguish Jenny that I have felt. I never used to shed tears Jenny, but now they cannot be kept back. When I think of our home in the South, the comfort we could have taken had I not bought so much, all would have been well. . . .

. . . I feel Jenny as if all rejoice at my failure in the South, but thank God I still have hands, and shall yet triumph. If you love me, then I can work hard and nobly, if not, I care not what becomes of me. I am sick and tired of human nature. Jenny don't you know, I have got too good a heart to get along in dealing with men. I wish it was in my nature to be mean. What I call mean, you understand. It would be better for me.[65]

Clearly, crop misfortunes were only the beginning of the Kings' woes in Alabama.

CHAPTER 8

The Battle of Life

January 1868–October 1903

——— ⚬⚭⚬ ———

During the dark days of the rebellion,
none done their duty better I trust, and to cap the climax,
I have the best wife that ever [a] mortal was blessed with.

James King's retreat from the South ended in Cincinnati, where he sought out his old comrade, Ephraim Hall. Hall was engaged in courtroom stenography and in teaching phonography—a phonetic system of shorthand. With the Kings' Southern adventure proven a fool's errand, James at the age of twenty-five again found himself seeking a means of making a living, and he hoped to leverage his blossoming talent for shorthand. Hall introduced James to Andrew J. Graham, whose brother Arthur had served in the 11th Michigan. Andrew was the author of *Graham's Handbook of Standard Phonography*, a book James had purchased with Hall's encouragement during the siege of Chattanooga. James now accompanied Graham to the district court and watched in amazement as the phonography master transcribed the judge's words flawlessly and with the greatest of ease. The mastery of shorthand required years of hard work. James at this time could write about eighty words per minute—a potentially marketable skill that did him credit—but a handful of stenographers nationwide could sustain triple that speed. Hall and Graham inspired James to continue devoting his every spare moment to shorthand studies.[1]

James was humbled by his financial situation and urgently needed to secure an income. Before the end of January he landed a clerkship in the office of Michigan's auditor general, William Humphrey, and moved to Lansing. Jenny and daughter May initially remained in Fabius. "I pray to the good one who watches over us," James wrote, "to hasten the day that will end our separation. Jenny I love you more than any man ever loved a woman, and I can hardly live without you." The clerkship, paying fifty dollars per month, was a stopgap solution to the Kings' money woes. James bided his time, keeping one eye on movements afoot in the state legislature to appoint stenographers to the courtrooms. "I suppose I can have a position here as long as the Republicans keep in power if I want it," he explained to Jenny, "but the wages are not enough."[2]

Additional funds trickled in from James's military service. He was awarded an invalid pension of six dollars per month for disability caused by his war wounds, and he finally received a long-awaited cash bounty for his service with the 11th Michigan. The ex-soldier visited his family whenever time and money allowed. His letters were brimming with comments about how he missed Jenny and May. By mid-April 1868 he was boarding at the home of Theodosia Branch in Lansing, and within one month of that, he took up phonography tutoring on the side. In early June he asked Jenny to forward his army discharge papers so he could proceed with initiation into the new and rapidly growing veteran's organization, the Grand Army of the Republic. James would ultimately rise to the GAR rank of adjutant general.[3]

He soon secured lodging sufficient for the whole family and was reunited with Jenny and May, who joined him in Lansing. Jenny read scores of books aloud to her husband to provide him with phonography practice. In January 1869 Jenny was midway through another pregnancy, and she again removed to Fabius, taking May with her. "Lonesomeness gnaws on me like a dog on a bone," James remarked soon after, but he kept busy. He tutored two phonography students and continued his own shorthand studies. He took stenographic proceedings notes for his GAR post and managed their bookkeeping, all in addition to his day job. In the weeks leading up to Memorial Day he penned an article, "Honor our Fallen Heroes," for the *Lansing Republican* newspaper. When the holiday came, the GAR post organized the strewing of flowers over the graves of war veterans, a project near and dear to James's heart.[4] He also assisted that year in the publication of the *Manual of the Grand Army of the Republic, Containing its Principles and Objects Together with Memorial Day in the Department of Michigan, May, 1869, List of Officers, Etc.*, edited and compiled by Senator Isaac Cravath. James was listed in the front matter as the sole point of contact for all orders and correspondence.

In the meantime, an old friend returned and helped stave off James's feelings of solitude. A letter arrived in April from Linus Squire, who had remained in Tennessee and served in the Freedmen's Bureau. His wife had passed away following an illness. Linus sold the gin that he and James had sweated over in

Thompson's Station, then returned to the Great Lakes State and by May 16 had moved in with James. "Linus and I are the worst old bachelors you ever saw," James told Jenny. Linus soon landed employment in the auditor general's office.[5]

As time passed, James's letters confided heightening fears over Jenny's health during pregnancy. Herbert Holbrook King was born on May 24, 1869, and the danger passed. The Kings named their firstborn son after Daniel Webster Holbrook. The family was back together in Lansing in early June, and their patriarch continued to toil for the auditor general. In early 1871 James's shorthand skills finally paid off when he recorded a speech delivered before the Michigan senate by Senator Cravath in advocacy for the creation of a state board of health. Proponents of the measure asked James to transcribe the speech and then submitted his work to the *Lansing Republican*, where it was duly published. The quality and aesthetics of James's phonography caught the eye of the newspaper's managing editor, publishing magnate William S. George, a seasoned newspaperman who had penned essays for William Lloyd Garrison's abolitionist newspaper, *The Liberator*, before the war. Shortly thereafter, George came across James on the streets of Lansing and said, "I have an idea that I could use a shorthand writer to advantage in dictation and would like to have you come to my house for a trial." James gladly accepted, and he took down the dictation of a column later published in the paper.[6]

One week later, James was walking the city streets when he again saw W. S. George, this time standing at the entrance to the newspaper office. "Mr. King," the newspaperman said, "if you will come with me I will give you the city editorship of the *Republican* and $1,000 the first year." The modest phonographer was stunned. "Why, Mr. George," James replied, "outside of two shorthand reports, an abbreviated report for the *Adrian Times*, and the dictation which you gave me, I have never written a line for a newspaper and know nothing about newspaper work." "You are just the man I have been looking for," George replied.[7] In that one joyous moment, the financial struggles, the wanderings, and the painful family separations in James King's life came to a welcome end. For the next fifteen years, he would make his home and career in the city of Lansing, Michigan.

The *Republican* was founded in 1855 to establish a much-needed mouthpiece in the state capital for the nascent Republican Party. The newspaper grew steadily in conjunction with its hometown and gained a reputation for capably expounding Republican principles. George had taken over as publisher on January 1, 1869, and later that year he had expanded the weekly paper to consist of eight seven-column pages. James was hired on as the paper's first city editor. George also served as Michigan's state printer, and being in his employ had its perks. The Michigan legislature passed a resolution in January 1873 awarding James a contract to compile and publish the latest edition of the enormous manual provided by the state to its politicians.[8]

Over the years, James seized every opportunity to socialize with his old army friends. At the 11th Michigan's 1873 reunion, speaking before a crowd of a hundred comrades and close to three thousand spectators, he said, "There is an attraction about everything connected with the war." He became the closest thing the regiment had to a unit historian. At the 11th's festive annual reunions, he delivered lengthy oratories that resurrected the turmoil and adventure of the 1860s. A newspaper reporter described one such reunion speech: James, "familiarly called Jimmie, and popular with everybody, read a patriotic, historical and eloquent sketch prepared by himself . . . [recounting] the battles the regiment participated in and giving personal reminiscences that brought forth cheers and laughter. Comrade King is well posted, is always interesting, was born a soldier and is today one of the most popular boys in the regiment."[9]

In September 1873 *Lansing Republican* political editor Stephen D. Bingham retired, and James filled his shoes while W. S. George served as news editor. For the next twelve years, James's editorial voice and opinions would dominate a major political organ of the Republican Party. George's counterpart at the *Allegan Journal*, Don C. Henderson, declared that during this time period "the *Republican* became standard authority in all political matters." On January 5, 1875, under James's guidance, the *Republican* made a successful leap to semi-weekly issue. Five years later, he successfully transitioned the newspaper to tri-weekly publication. That year, a county historian wrote of James and the *Republican* that "no paper in Michigan is more frequently quoted, and none is more rarely caught in mistakes of fact. . . . James W. King has developed fine talents for newspaper-work of every kind, and is an accomplished short-hand writer." One of James's industry peers later stated that "we could always rely upon any statement or figures given in the Lansing Republican while Jimmie King was editor."[10]

In the meantime, the King family had continued to grow. James Guy King was born December 18, 1870, John Willard on October 28, 1872, and Henry Burr, on April 24, 1876.

W. S. George passed away on December 27, 1881, at age fifty-six. James stood up before the Michigan Press Association and delivered a touching tribute to the man who had changed his life. "He made confidants of his assistants," James said, "and was ever ready to heed and accept counsel from them. . . . An employee who was capable, honest, and industrious was never deserted by Wm. S. George. . . . Those who knew him best appreciated and loved him most." In

Left: Jennie May King. (Author's collection)

Below: The Kings at their house on Townsend Street in Lansing circa 1876, during James's editorship with the Lansing Republican. James, bearded, leans on the fence, with May facing him. Jenny stands nearby, on the far side of the fence. The child in the high chair on the porch is probably Henry Burr King. The other two adults are Jenny's younger sister Mary Esther Babcock and her soon-to-be husband, George Avery. One of the four King sons is missing from the photo. (Author's collection)

George's absence, James rose to the lofty post of chief editor. He had his finger on the pulse of national politics, and his name became known in Washington. Surviving correspondence shows that he was in contact with such individuals as Congressmen Oliver Lyman Spaulding and Byron Mac Cutcheon.[11]

James King the editor took stands on a number of issues. The *Republican* under his tenure was bitterly outspoken about continued voting irregularities in the South (complaining, for example, of "the kukluxed electoral vote of six southern states" in the election of 1884) and the pandemic, ongoing abuse of blacks in Dixie. James's editorials spoke favorably of black suffrage—something he had adamantly opposed back in his cotton-raising days—and asserted the full equality of the races: "God made the black man as well as the white.... One man is just as good as another in this free America." The paper enthusiastically reported a visit to Lansing by Sojourner Truth, "this remarkable colored woman and prophetess of her race." Harriet Beecher Stowe's writings were said to "elevate the tone of humanity," and *Uncle Tom's Cabin* was rated second only to the Bible in its influence for the betterment of mankind. The *Republican* still referred to Democratic Southerners as Confederates, and Dixie was "all rebeldom." Jefferson Davis was derided as "that hoop skirted relic of the great rebellion," a nod to rumors that the Confederate president had attempted to escape capture disguised as a woman.[12] The war was long over, but the political chasm between North and South would take decades to heal.

Unsurprisingly, temperance was a pet project of the ex-quartermaster sergeant who had so frowned upon his comrades' drinking in wartime. James suggested the application of high taxes on alcohol to achieve the desired result. His editorials blamed stiff drink for the lion's share of society's ills. Free trade was scorned in the *Republican* as injurious to Michigan laborers, and examples were advanced to prove the advantages of protective tariffs. Unions, with their productivity-stifling strikes, were frowned upon. Overall, the paper toed the party line on most major issues. The Greenback Party was vilified, and opposing newspapers such as the *Detroit Free Press* were bluntly derided for their editorial opinions—common practice in nineteenth-century journalism. "Democrats are not all rascals," one *Republican* editorial proclaimed, "but most of the official rascality is democratic."[13]

Tidbits about the Kings' friends and family occasionally made the news. Visits to Lansing by Jenny's parents, Linus Squire, and Reverend Luther Trowbridge were deemed newsworthy. A paragraph was devoted to a skating accident that left son John Willard King unconscious for an hour, and Theodosia Branch's trip to Detroit to replenish her millinery stock was publicized. Of more importance to James was the untimely death of Ephraim Gaylord

Hall, the subject of a detailed obituary in the edition of April 14, 1881. Hall had pursued a legal career, and later went on to excel as editor of the *Cincinnati Gazette*. Reports from Cincinnati and publications out east ascribed Hall's demise to overwork, but James declared that Hall's "old comrades know that the severe wound which he received [at Stones River] so close to the base of the brain, and the subsequent torture which he endured at Libby prison, gave a shock to his nervous system from which he never rallied."[14]

James's shorthand skills served him well on the job. A biographical album said of his newspaper days,

> There was scarcely an hour but what he hurried matters by his skillful knowledge of shorthand. It was useful in a thousand and one different ways. His reporting covered testimony in legislative investigations, political speeches, convention and institute proceedings, and lectures of all descriptions and characters. This work included the utterances of many of the most noted men and women of the land. Among those who have complimented him personally for accuracy in speech reporting, of his own State, were Senators Chandler, Ferry, Christiancy and Palmer; Governors Croswell, Jerome and Alger; Congressman Horr, and scores of others.

James's capabilities are illustrated by an event that occurred near the end of his editorship with the *Republican*. Benjamin Butler delivered a campaign speech in Lansing one evening during his unsuccessful 1884 presidential campaign. James, as the *Republican's* chief editor, went to cover the event, only to find an impenetrable crowd barring his approach to the speaker. He acquired entry to a nearby building, crawled out a second-story window, and with his legs dangling from the ledge, took down the entire speech in darkness through the power of shorthand, unable to see his pencil marks or even to make out the lines on the page. The speech was successfully published in its entirety the next day.[15]

An advertisement for the *Republican* in the 1883–84 *Lansing City and Ingham County Directory* provides insight into how the paper marketed itself in James's day:

> It is trustworthy in every respect.
> Editorials on the current events of the day....
> A concise and readable abstract of Congressional proceedings, when that body is in session.
> A careful condensation of the important news, by telegraph, from the whole world....

The tone of the Republican is suitable for the family circle in every respect. Its aim is to inform, reform and elevate the people.
Its principles are firmly and radically Republican, but it never denies a fair hearing to an opponent.[16]

The newspaper came under new ownership at the beginning of 1886. James's journalism career came to an end, apparently by choice, as he wished to return home to care for his aging parents.[17] In his final column, published January 2, 1886, he addressed the readership regarding his departure.

During our newspaper service hard blows have been struck and taken in return, but never from a personal standpoint, and retirement from the editorial chair is accompanied with the belief that I have no personal enemies. . . .

. . . The Republican has ever battled squarely for what it believed to be right, without stooping to personalities, or striving to advance the interests of personal favorites. . . .

And so I see that change is stamped on everything. For nearly 18 years I have been a resident of the beautiful city of Lansing. Within her borders, from all classes, a host of friends rise before me. To them and to the friends scattered in every portion of the state, my last editorial act is to say good-by and God speed.[18]

The return to Fabius was well timed, for Benjamin and Eliza King hadn't long to live. Eliza passed away on March 15, and Benjamin followed his wife exactly six months later. Benjamin was remembered in a biographical album as "a very early settler and prominent man in this State." He had served as a vice president of the St. Joseph County Pioneer Society and was a member of the Pioneer Society of the State of Michigan. James penned his father's obituary for the latter organization: "Mr. [Benjamin] King, for many years, took a deep interest in public affairs. He was an active participant in the antislavery struggle, and during the war of the rebellion the Union had no firmer friend than he. He fought the battle of life like a true soldier, and well performed his part as a husband, father, and citizen."[19]

James now resided in his childhood home—the same house he had helped Charles Rice construct before the war—and represented the last of the Benjamin King family in Fabius. His brother Henry had passed away in 1872, surviving only into his early forties. John lived in Waukegan, Illinois. Rebecca moved to Rockford, Illinois, after marrying Eliza King's nephew Albert Dickinson, and

James M. Whallon.
(Belknap, *History of the Michigan Organizations at Chickamauga, Chattanooga, and Missionary Ridge*)

sister Eunice had married Charles Howe and ventured off to the frontier settlement of Raymond, Dakota Territory.[20]

Shorthand again proved the essential driving force behind James's career. Governor Cyrus G. Luce appointed the former newspaper editor as stenographer of the Fifteenth Judicial Circuit on October 7, 1887. The Fifteenth Circuit then served the counties of St. Joseph and Branch.[21] This job proved every bit as stable as the editorship with the *Republican*; in fact, James would serve out the rest of his days in this same employment.

Somehow, the war always found its way back into James's life. In 1893 he was heavily involved in preparations for the opening of the Chickamauga and Chattanooga National Military Park. He composed detailed articles for Michigan newspapers regarding the park preparations. He traveled to the Chickamauga battlefield and worked with his old comrades James Whallon and William Whitney to stake out the appropriate location for a monument to the 11th Michigan on Horseshoe Ridge. The military park was dedicated on September 18–20, 1895, the thirty-second anniversary of the battle. On September 18 Whallon, Whitney, and James were joined by Borden Hicks, John Underwood, and a dozen other veterans of the 11th. The aging ex-soldiers gathered on Horseshoe to dedicate the completed monument, which was crowned with a statue of William Stoughton, who had passed away seven years before. Forty veterans of the Nineteenth Illinois happened to arrive at that moment, and an enthusiastic, impromptu brigade reunion ensued. The two regiments agreed to hold their dedicatory services together. Michigan secretary of state Washington Gardner opened with a prayer service, and James followed with an eloquent speech that conjured up scenes of one of the deadliest battles in American history. He was proud that the Nineteenth Illinois had nominated him as their speaker of the day as well.[22]

Time marched on, delivering its accustomed share of joys and sorrows. On March 2, 1894, the Kings suffered a traumatic loss: their youngest child, Eva Fay, succumbed to scarlet fever at the age of two years and three months.[23] The toddler was buried with her paternal grandparents at Riverside Cemetery in Three Rivers.

The monument to the 11th Michigan on Horseshoe Ridge, photographed soon after its dedication. Stoughton's statue had its arms broken off several years ago. (Belknap, *History of the Michigan Organizations at Chickamauga, Chattanooga, and Missionary Ridge*)

James never got over his war injuries. For the rest of his life, he was tormented by two debilitating wounds. His right arm could endure little exertion, and the associated discomfort worsened whenever it rained. Yet the shell wound to his left shoulder proved much the worse of his war injuries. For years, Jenny applied heat and medicines to his bad shoulder in a daily attempt to ease the suffering. Time did not fully heal these wounds; James's pension records indicate that his distress in some ways increased over the years. The shoulder forever remained a source of pain, and the pension file is replete with testimony stating that the ex-soldier suffered from severe heart palpitations—sometimes leaving him bedridden for days at a time—that worsened as the years passed, and that dated from his wounding near Atlanta. In April 1895 James lost consciousness while in the courtroom, and the doctors despaired of his life. He was diagnosed with hypertrophy and dilatation of the heart. For the next eight and a half years, hardly a month passed without the patient consulting his doctor, Lawrence Knowles, about cardiac troubles.[24] Yet he remained an active man.

In 1897 he joined Ashbel Snyder—brother of the 11th Michigan's Sergeant Major Washington Irving Snyder, who had captured Confederate brigadier general Daniel Adams at Chickamauga—in returning the general's sword

An 1894 photo, likely taken at Eva Fay King's funeral. James and Jenny are together at top right, with James holding a young child. May and her husband are seated in the middle row just right of the pillar, with their daughter Lillian in May's lap. Jenny's parents are in the same row, just left of center. Herbert Holbrook King is seated before the pillar, bottom center. The two mustached men seated a short distance to either side of Herbert Holbrook are his brothers. (Author's collection)

to his sons in New Orleans. Adams's family in turn donated the blade to the Louisiana Historical Association.[25]

When America went to war with Spain in 1898, James and Jenny's fourth child, twenty-five-year-old John Willard King, enlisted as a sergeant in Company K of the 33rd Michigan Volunteer Infantry. The war with Spain was a whirlwind affair. Will, as his family called him—perhaps a nod to Will Iddings—departed Michigan in late May, and by the first of July he had trained in Virginia, sailed off to Cuba, and experienced war in all its fury. On July 1 his regiment participated in the Battle of the Aguadores, a feint in support of the famous assault on San Juan Hill. Will was slightly wounded by a shell but returned home safe and was promoted to second lieutenant.[26]

In that same year of 1898 James served on the Organizing Committee of the National Stenographers' Association. Around the same time, he was elected president of the Law Stenographers' Association of Michigan, and he would secure reelection each successive year until his death.[27]

By 1900 Jenny's aging parents had moved in with the Kings. Darius was now eighty years old; Ruth was seventy-seven. The King children had all moved out and made lives of their own. May had been married to Adam Shafer for fourteen years and still resided in Fabius. As of October 1903 Herbert Holbrook King was secretary of the Three Rivers Foundry; James Guy was a mail clerk; John Willard was a bookkeeper with the Three Rivers Paper Company; and Henry Burr had followed in his father's footsteps, serving as a courtroom stenographer in Chicago. In addition to the old King farm and house, James and Jenny owned another 140-acre farm in Fabius, two lots on beautiful Corey Lake (one graced with a lakeside cottage), and a house on Townsend Street in Lansing.[28]

In his twilight years, James spent more and more time reflecting on the Civil War, the defining period of his life. Throughout the postwar decades, the 11th Michigan's quartermaster sergeant spilled a considerable volume of ink sharing his indelible recollections of the rebellion. His discourses were published in various Michigan newspapers, and frequent comments about the war appeared in the *Republican* during his editorial tenure. Variations of his autobiographical sketch were published in several periodicals and in at least one book, *Portrait and Biographical Album of St. Joseph County, Michigan*. The May 1, 1896, issue of the *Three Rivers Tribune* carried James's history of the Battle of Missionary Ridge, and on January 11, 1901, the *Republican* printed his "11th Michigan Infantry," a paper he read before the Charles T. Foster GAR post in Lansing. This extensive speech showcased his history of the regiment, the twenty-seven-page typed draft of which still survives.[29]

Two of the seminal accounts of Michigan's role in the Civil War are John Robertson's thousand-page opus *Michigan in the War*, first published in 1880, and Charles E. Belknap's *History of the Michigan Organizations at Chickamauga, Chattanooga, and Missionary Ridge*, which hit the presses in 1897. Both works were authorized by the state legislature. The former tome, in its coverage of the 11th Michigan's actions at Missionary Ridge, offered brief but powerful praise for James's part in the battle: "Colonel Stoughton says of Quartermaster Sergeant Jas. W. King, of the 11th: 'Taking a voluntary part in the battle, he behaved with conspicuous gallantry, and received a severe wound.' It has been very satisfactorily established that Sergeant King was among the first to reach the summit." Belknap's work, in addition to repeating the above comments almost verbatim, published James's account of his own actions in the battle (quoted in chapter 5).[30]

A wedding, circa 1897. James stands up front toward the right, holding a large hat at his chest, with Jenny standing to his left (viewer's right) and Jennie May on his other side. The groom, at center, is probably either John Willard King or James's nephew (Henry King's son), James Arthur King. Jenny's parents stand a short distance left from the groom, with Darius sporting his gray beard and Ruth donning a bonnet. (Author's collection)

James King's comrades immediately recognized, and never forgot, his gallant charge at Missionary Ridge. In the immediate aftermath of the battle, tributes to his bravery, penned by Edward Frost and Borden Hicks, graced the pages of the *Three Rivers Reporter*. James King, wrote Frost, "voluntarily took a musket and went into the ranks of Co. E. He did yeoman service, but his patriotism cost him a broken arm." Hicks, under whose command James committed his acts of heroism, added, "There is not a braver boy in the whole army than he. His position did not require him to take a gun and go into battle, but he volunteered to do it." Over the ensuing decades, James's army friends encouraged him to seek a Medal of Honor. James did not covet such recognition in his youth, but toward the end of his life he experienced a change of heart, and in this he was not alone. There was a resurgence of interest in the medal in the 1890s, likely triggered in part by the publication of several books about the decoration. Well over 600 Union soldiers received the medal in that decade alone—including the 11th Michigan's William Whitney, who braved Confederate sharpshooting to gather desperately needed ammunition from dead and wounded Rebels during a lull in the assault on Horseshoe Ridge.[31]

James was nominated for the medal by Senator Julius Caesar Burrows, a former captain of the 17th Michigan who had enjoyed some past success in obtaining medals for his constituents. Burrows recommended James in a letter to Secretary of War Elihu Root, stating in part, "It is believed that he is entitled to a medal for special gallantry. . . . Mr. King is numbered among our best citizens in Michigan, and it will be specially gratifying to me, if he be found worthy." But Burrows acted on November 26, 1901, shortly after Root convened a board of officers to review, reform, and stiffen the medal's standards. The timing could hardly have been worse, and the nomination was submitted with only one deposition—penned by Captain Whitney, and strongly complimentary of James's bravery, but somewhat lacking in specifics about his actions in the battle. Whitney's testimony alone may well have proven sufficient in the 1890s, but 1901 was a different story.[32] Nonetheless, James was caught by surprise when he received a rejection letter from Assistant Secretary of War William Cary Sanger. He drafted a response to Senator Burrows.

———

William G. Whitney, earning the Medal of Honor on Horseshoe Ridge. (Beyer and Keydel, *Deeds of Valor*)

Elihu Root. (Library of Congress)

Julius Caesar Burrows. (Library of Congress)

Three Rivers, Michigan,
February 17, 1902

Hon. J. C. Burrows
Dear Friend and Comrade:

I thank you for the efforts you have made in trying to secure for me a national Medal of Honor.

Years ago, many of my comrades urged me to make application, but then I did not look upon the matter as I do at the present time. It ought to have been done before Gen. Thomas, Gen. Stoughton and scores of others of my brave comrades had passed to the silent shore.

Undoubtedly the honorable assistant secretary of war believes he has followed the law and precedent; but to one who for four years witnessed the brave acts of the bravest soldiers that ever fought on [a] battlefield; who for fifteen years as a shorthand reporter and editor attended and reported Grand Army encampments and soldier gatherings; who for the last seventeen years has spent his whole time in courts of justice where cases are decided according to law and evidence; who was selected not only by his own regiment but also by the 19th Ill. Infantry to deliver the address at their late dedicatory services held on the battlefields of Chickamauga and Mission Ridge, the finding not only seems wrong, but it is unjust and cruel.

The assistant secretary says: "The conduct of Mr. King on the occasion that is made the basis of the application for the award of a medal is shown by the testimony to have been gallant and highly creditable to him, and his entire record as a soldier is one of which he and his friends may justly be very proud, but his conduct on this particular occasion was not of that *most distinguished character* for which, alone, the medal can now be awarded." The italics are mine.

Wm. G. Whitney, who testified in my behalf, as brave a man as ever wore the Union blue, who holds a national medal for conspicuous gallantry at Chickamauga, and whose affidavit is now before the War Department, among other things, says: "I was in active service for four years, and while in such service witnessed many acts of bravery on the part of soldiers, and this brave charge of said King, outside the line of his duty, and knowing the great danger he was facing, was one of the bravest acts that ever fell under notice of deponent." Which is the better authority on *most distinguished character* [emphasis in original], the assistant secretary of war, clothed with a little brief authority, who may not have been within a thousand miles of Mission Ridge; or Wm. G. Whitney, the brave Gen. Stoughton, and the other gallant comrades who passed with me through that storm of musket balls, shell, and canister, where the very hell of battle centered at Mission Ridge?

This is the first time I have ever argued a case in my own behalf, but your kindness in this and other matters of the past years gives you a right to know, more in detail, the bare facts on which I based my application for a medal. The night of the 24th of November saw Hooker's men in possession of Lookout Mountain. Sherman's forces were crossing the Tennessee to gain lodgment on the north end of Mission Ridge. I knew that the struggle on the morrow would be for the possession of that mountain crest. Major B. G. Bennett, I think you knew him personally, a gallant, warm-hearted, impulsive man, was in command of the regiment. As Q.M. Sergeant, my duties were with the stores, and before I could go to battle with my comrades, permission must be obtained from the commanding officer. I knew Major Bennett would refuse the asked for permission. Gen. Stoughton was in command of [the] brigade. I sought him at brigade headquarters after dark. At first he refused to grant my request, but after importuning him for some time, he finally consented. I went to camp and found Major Bennett in front of his tent. I told him what I had done, and he said: "James, you are a little fool. A lot of us will die tomorrow on that ridge, I shall not come out of the battle alive. It is not in the line of your duty, and I beg of you not to go." I replied that it was no worse for me to go into battle and risk my life than it was for my comrades, who had not been quite as fortunate as I in getting a position where they could escape bullets. The eventful morrow came, and with it, under the eye of the immortal Grant, the grandest charge ever recorded in military history. When the final order was given, at four o'clock, for the Army of the Cumberland to charge, we crossed an open field for a half a mile swept with musket balls and shell, and I was among the first to scale the Confederate works at the foot of the ridge. Remaining in the works for perhaps two or three minutes, we started for the crest of the mountain three-quarters of a mile away, at an angle of more than 45 degrees. The top of the ridge was a blaze of fire from artillery and musketry, and showers of canister and musket balls swept the steep ascent. Comrades and history have given me the credit of being one of the first to reach the mountain crest. About 100 of the men of the 11th and a detachment of the 19th Illinois were with me when the Confederates in our front gave way and sought safety in flight down the eastern slope of the ridge. Just to our right the Confederate line was intact, and not more than five or six rods away was a semi-circular earthwork where [there] was a battery of six guns which were being worked on the Union troops farther down the ridge. There was also a support of four or five hundred Confederate infantry, who kept up a murderous fire on our men below. A chestnut tree about 20 inches in diameter stood on the crest of the ridge, where it began to slope to the east, and a log nearly as large was lying beside it. The two formed a niche, and I took advantage of the safety it afforded. Sheltered as I was, I would load my rifle while lying on my back, then turn over and fire across the log at the Confederates who were still working the guns. Every time I showed myself above the log, Minie balls would whistle over me or strike in the log at my right. I had loaded my gun for the 21st time, and on rolling over to cock and cap it, I raised my right arm a little too high, and a Minie ball struck it, breaking and crushing the bone just above the elbow joint. I could do no more fighting on my own account, and at the risk of my life, I jumped and ran back about a rod to a large tree, behind which was Byron D. Liddle of Co. D. He had just fired his gun and I told him to cap mine and give them a shot. We had made it so uncomfortable for the Confederates from our flanking fire that the infantry support and artillery men fled precipitately down the eastern slope of the ridge. I then had Comrade Liddle tie a handkerchief around my arm to prevent loss of blood, and then went back down the slope perhaps fifteen rods to the main body of the regiment to tell Major Bennett we had broken the Confederate line on top of the ridge, that the enemy had deserted their guns, but that they might return, and that we would need help to hold them. The major had been killed about halfway up the ridge, and Capt. P. H. Keegan was in command. I hurriedly explained the situation to him, helped to hurry the boys up, and with them again returned to the top of the ridge. Remember that when I did this, my arm was mashed and dangling by my side. When I reached the top of the ridge the second time, the summit was swarming with troops, and there was no danger that the enemy could repossess their former ground. Men were almost crazed with joy at the unexpected turn things had taken, and at the grand and heroic work the Army of the Cumberland had just performed.

I remained on the top of the mountain until I saw the last of the Confederates disappear from view, with my comrades in close pursuit, and then, between sundown and dark, I put my Springfield rifle on my left shoulder, passed down the cragged side of the mountain, and walked three miles to our camp on Cameron Hill. Here I left my gun and accoutrements and walked a distance of a mile down in the city to [the] general hospital. When they got ready to attend to my wants, they laid me on the table to cut my right arm off. There were 13 surgeons in the counsel and all said it must come off except my regimental surgeon, Dr. Elliott. Through his intercession it was saved.

It is my opinion that if the present assistant secretary of war had been obliged to take that four-mile walk, under the same conditions that I did, to say nothing of the voluntary risk of life in battle, he would have thought that his conduct was of such a *most distinguished character* [emphasis in original] that it would have taken more than a bushel of medals to fully compensate him for his bodily sufferings, and mental anguish caused by the expectancy of losing his good right arm. Linus T. Squire, the adjutant of my regiment, as cool and gallant a soldier as it was ever my good fortune to serve with, is now a clerk in the sixth auditor's office at Washington, and knows every detail of the matters I have above stated.[33] He was shoulder to shoulder with me in that charge until nearly two-thirds of the way up the mountain side, when he was disabled.

I submit that the recommend on the back of my old discharge, signed by my brigade commander, and nearly all of the officers of the 11th Mich. Inf., at muster out, that alone ought to entitle me to a medal. I say this with all due deference to the contrary opinion of the hon. assistant secretary of war.

The breaking of my left shoulder at the siege of Atlanta the following summer, by a piece of shell, weight a pound and a half, of course cannot be taken into account when we are seeking to establish that "most distinguished character for which alone the medal can now be awarded."

I ask your indulgence for so frequent a use of the pronoun I, but there was no other way to do justice to the case. Hoping that I have not wearied you by so extended a statement, I remain as ever your friend and comrade, J. W. King.

Neither James nor Senator Burrows gave up easily. They recognized that Whitney's lone deposition hadn't done James's actions justice, so the nominee garnered half a dozen additional affidavits during March and April 1902. The deponents included Sergeant James Bouton of Company G; Private Elliott Gray, Company C; Private James Haines, Company E; Corporal Simeon D. Long, Company D; Private Hiram D. Wescott, Company A; and finally, Captain Borden Hicks. The new testimony was powerful and clearly placed the sergeant—voluntarily engaged in a desperate battle he did not belong in—charging ahead of his brigade against fortified positions not once but twice. All six soldiers swore under oath that they had never seen "a braver act or a more heroic deed" than James committed at Missionary Ridge. Burrows forwarded this stack of depositions to the assistant secretary of war on December 2, 1902, "with the hope that you may see your way clear to make the award." But 1903 was destined to become the first year since 1882 to pass without the awarding of a single medal to a Civil War veteran. The War Department let James's paperwork gather dust while it focused on halting the flow of medals for Civil War action once and for all—with the backing now of the attorney general and the president. The department published a circular on March 16, 1903, requiring that all nominations must be submitted while the soldier was still on active duty. No exception was allowed for applications already received, so this dictate retroactively terminated James King's Medal of Honor nomination. The full merits of his case would not be considered. The War Department did not reply to Burrows until October 13, 1903, and as it turned out, James would not be around to hear the final verdict. On April 23, 1904, Congress overrode the discharge date rule, but for the 11th Michigan's quartermaster sergeant, the correction came too late.[34] Jenny had been left a widow.

James's medical condition had continued to deteriorate throughout 1903. He had confided in his comrades at recent regimental reunions that he believed his heart condition would end his life. In the summer of 1903 his health entered an irrecoverable downward spiral. In August he told Isaac Kemberling that his physical state was worsening, and that he did not expect to live long. On September 8 he suffered an attack of such severity that his doctor wrote, "It would seem that he hardly could survive so bad was the arrhythmia."[35] But the family needed money, and the ailing ex-soldier continued to toil.

October 9, 1903, was a day like any other at the courthouse in Coldwater. James and Jenny had celebrated their thirty-eighth wedding anniversary just three days prior. It was around 3:00 P.M. and circuit court was in session. James was seated in the courtroom between judge and jury, finishing up some case-work for Judge George Lewis Yaple. He set his pen aside for the moment. The prosecution initiated its closing argument. James's presence was not required. He rose to his feet and walked to the attorneys' room, where he habitually entertained the lawyers with clever stories in what one of them termed "his usual jolly frame of mind." The popular stenographer had a reputation for sharing his knowledge of the court system with the younger attorneys and had been granted honorary membership in the Branch County Bar Association. He took a break for a minute or two before heading back toward his desk. As he reentered the courtroom, James suddenly reeled and reached for the sheriff's desk to steady himself. A moment later, all life left his body. The sixty-one-year-old collapsed and struck the floor as dead weight. A commotion broke out as several people rushed to his still form; others ran to seek a doctor. The lifeless body was lifted onto a table and physicians arrived promptly, but there was nothing to be done.[36]

The outpouring of grief in the community was immediate. The *News Reporter* of Three Rivers carried a story entitled "James W. King: A Sketch of the Prominent and Well Known Man," which began, "The sudden news of the death of the above named gentleman was received in the city Friday afternoon and has cast a gloom over the city and entire community for he was well known and beloved by all who knew him." The paper presented a long list of touching

James Wood King, late in life. (Author's collection)

and admiring epitaphs from friends and colleagues. Judge Russel Ralph Pealer remembered James as "a loyal friend." Former congressman Judge Yaple stated that "Mr. King has been court stenographer in the circuit for ten years, and our relations have of course been very close. In all that time I never heard him make a discourteous or disparaging remark about any person. I never heard him criticize a human being." Mrs. B. E. Andrews declared the deceased a man "of great purity of life and character, of irreproachable habits and a model husband and father. In my lifelong acquaintance with him I have never met anyone who had not the greatest respect for his sterling honesty." George Arnold described James as "broad minded with a magnetic influence which made him a leader among men. He had a host of friends who always appreciated a few hours spent in his society. He was one of the few men of whom we can say 'that the world is better because of their presence in it.'"[37]

Members of the bar association motioned to attend the funeral as a group, but railroad schedules prevented coordination on so grand a scale. Mayor Milo Campbell of Coldwater declared, "I think there isn't a member of this bar but that wants to show every possible respect to 'Jim King' as we all knew him. I think we all loved him." The funeral was heavily attended, and the floral arrangements provoked a comment of admiration in the local paper. Judge Yaple performed the eulogy. James was laid to rest among the graves of his parents, daughter Eva, and brother Henry. Nearly two decades would pass before Jenny joined him.[38]

Henry Humphrey, who had worked at the desk adjacent to James at the auditor general's office, was also his neighbor during his days with the *Lansing Republican*.[39] Humphrey wrote in grief to the *Republican*,

> I wish I could express in words just the impressions I have about the life that went out so suddenly and so painlessly yesterday afternoon. The friendship, co-partnership, into which we entered thirty-three years ago has never been dissolved....
>
> ... How can I speak of his large-heartedness, his tenderness and his loyalty to friends. As far as I know he was not a member of any association of organized goodness. His kind acts were entirely unorganized, spontaneous, offhand and genuine....
>
> I am wondering now, as many another one may be, if my friend still thinks? If he thinks of home? If he thinks of others? If he thinks of me? And if he does, what are his thoughts; how may he look and where may he be? For if any like he does think, does live, does watch us here, I think his capacity for a better life is abundant and his title to it clear.[40]

The grave of James Wood King at Riverside Cemetery in Three Rivers. (Photo by the author)

The grave of Sarah Jane King at Riverside Cemetery in Three Rivers. (Photo by the author)

The Michiganders who marched off to war with James King forever remembered the caliber of man and soldier he was. "Your life," Lieutenant Colonel Melvin Mudge wrote to James in 1901, "has been an active and useful one, and how can we know the good that springs from our deeds? Your life and works has been among the people, and you have had the eye and ear of the public, and your work has been well and faithfully done." After James's death, his comrade Lindley Harkness penned a deposition to help Jenny secure a widow's pension. Harkness dutifully provided the required statements of fact, but in closing he could not resist interjecting a comment outside the scope of his testimony. James King, Lindley Harkness wrote, "was as brave and true a man and soldier as ever drew breath."[41]

James W. King and the Medal of Honor

James King's conspicuous gallantry at the Battle of Missionary Ridge led to his Medal of Honor nomination in 1901, thirty-eight years after his acts of heroism, and offers a window into an infrequently discussed phase in the history of the U.S. military award system. Much ink has been spilled about the lax standards for awarding the Medal of Honor in the early decades of its existence, but much less known is the swing of the pendulum in the opposite direction at the turn of the twentieth century. James King's ill-fated medal nomination faced hurdles that simply did not exist for the more than six hundred Civil War soldiers who were awarded the medal in the 1890s.

Should James W. King have received the Medal of Honor for his actions at Missionary Ridge? During the Civil War—and into the early years of the twentieth century, when King was nominated—the medal represented the only decoration provided by the federal government for recognizing valor, and the criteria for earning the award were less demanding than at present. The basis of today's medal system is a hierarchy of awards known as the Pyramid of Honor, recognizing varying degrees of bravery and service, with the Medal of Honor at the pinnacle; correspondingly, the requirements have evolved to the point where Medal of Honor recipients usually perish in the act of earning the award. For the Civil War, however, it was quite the opposite case: less than 1 percent of medal winners died in the act of earning the decoration. It was primarily an award for survivors. In practice, a deed performed outside the line of duty and at mortal risk was likely to be deemed medal worthy.[1]

At Missionary Ridge, James fearlessly risked his life when his duty would have placed him literally miles away from danger. He sprinted ahead of his brigade in two separate charges—with the first concluding in hand-to-hand combat—under heavy artillery and rifle fire, and was among the first Union soldiers to crest the ridge. After engaging the enemy with short-range flanking fire at a critical time and place, he received a serious wound and still further exposed himself in an effort to hurry his comrades up the slope. Did these actions warrant a medal? The ranks of Civil War medal winners provide ample opportunity for comparison; one may readily identify a handful of individuals who voluntarily entered battle outside their line of duty and earned the award. Consider the case of regimental quartermaster Charles Murphy of the 38th New York, who grabbed a rifle at First Bull Run, fought with his regiment, and stayed behind to help the wounded after the battle ended—with the Rebels in control of the field and despite repeated urgings to retreat. Murphy was captured. Next, Private Henry T. Johns of the 49th Massachusetts, detailed as a quartermaster's clerk, voluntarily joined a picked squad of men to charge across open field under heavy fire at Port Hudson on May 27, 1863; the attacking force retreated after taking heavy losses, but Johns escaped unharmed. Sergeant John H. Cook of the 119th Illinois, detailed as a headquarters clerk, armed himself and joined his regiment at Pleasant Hill, Louisiana, on April 9, 1864. Cook deployed his company as skirmishers and urged his men forward into the fight under brisk rifle fire. He made it through uninjured. For our final comparison, we need look no farther than Missionary Ridge, just hours prior to the charge of the 11th Michigan. Seventeen-year-old drummer boy John S. Kountz of the 37th Ohio set aside his drum and—defying his colonel—took up a rifle to join in a charge during Sherman's assault on the Confederate right. The Ohioans were twice driven back with loss; on the second occasion Kountz took a bullet through his left thigh, falling closest to the enemy works out of his regiment's forty-one casualties. He was rescued by a brave comrade who went back for him, but he still suffered amputation. Kountz was found guilty of disobeying orders, yet his action became the subject of a dramatic poem, "The Drummer Boy of Mission Ridge"; he went on to be elected commander in chief of the GAR in 1884—and like the rest of these soldiers, he received the Medal of Honor.[2]

The judgment of valor is subjective, but it seems reasonable to say that James King's actions compare favorably with those of the medal winners described above. There was just one key difference between James's case and those of Murphy, Johns, Cook, and Kountz: James let the matter rest a little longer. The other four soldiers received their medals in the 1890s. By 1901, when the name James W. King was submitted for award consideration, the well had run dry. Secretary of War Elihu Root had become exasperated with the demands placed on his department to investigate medal requests for actions that occurred decades in the past. This, combined with concern over the lax medal standards of the 1890s, drove Root to overcompensate—the outcome of an honest desire to defend the sanctity of the decoration. The rate of application approval plummeted. With no lesser awards to fall back on, the medal decisions were necessarily all or nothing—and nothing became the comfortable default.[3]

Assistant Secretary of War William Cary Sanger's letter rebuffing James's nomination strongly implied that he was rejected for failure to reach a new and stricter standard: "that most distinguished character for which, alone, the medal can now be awarded." Sanger was alluding to the impact of efforts to tighten the award's requirements in 1897, even before Elihu Root's term as secretary of war. The army's Medal of Honor had been instituted by law in 1862 to honor those who "most distinguish themselves by gallantry in action, and other soldier-like qualities." But in June 1897 the War Department, at President McKinley's behest, altered this definition to reward only service "performed in action of such a conspicuous character as to clearly distinguish the man for gallantry and intrepidity above his comrades—service that involved *extreme jeopardy of life* [emphasis added] or the performance of extraordinarily hazardous duty." Suddenly, in the wake of this new stipulation, most medal nominations were denied. 1897 saw the issuance of 112 medals for Civil War action, but in the years to follow the numbers dropped off precipitously: 58 in 1898; 27 in 1899; 10 in 1900. Root's policies merely accelerated that trend. During the fiscal year ending June 30, 1901, 60 nominations were submitted and only 4 medals awarded— and even that rate of acceptance would soon appear liberal in retrospect.[4] For James King, in the end, it was purely a matter of bad timing. The subsequent publication of the War Department's March 1903 circular, requiring a soldier to be on active duty at the time of his nomination, terminated his attempt to appeal, along with any hopes for the case to be considered on its full merits.

James King's Medal of Honor nomination file is preserved at the National Archives in Washington, D.C. It is located in Record Group 94, Entry 501, Record and Pension Office File, document number 670252. The file contains seven affidavits. Those of Captain William G. Whitney and Corporal Simeon D. Long are presented below. The balance of the affidavits are worded very similarly to Long's. The other deponents were Sergeant James Bouton of Company G, Private Elliott Gray of Company C, Private James Haines of Company E, Captain Borden Mills Hicks of Company E, and Private Hiram D. Wescott of Company A.

William G. Whitney ... being first duly sworn, deposes and says: that he was late Captain of Co. "B" 11th. Mich. Volunteer Infantry in [the] late war of the rebellion; that he was well and personally acquainted with James King, late Quarter Master Sergeant of said Regiment and who now resides at Three Rivers, Mich; that said King was while in the service a young man of exemplary habits and was a good soldier; that since he was discharged from service he has continued to be of the same character and is a good citizen; that said Regiment was engaged in action at battle of Missionary Ridge Nov. 25th, 1863, and that deponent was then a Sergeant of said Co. and took command of said Co. on that occasion on account of losing his superior officers; that at said Battle, said King, who was at that time Q.M. Sergt., laid aside his duties as such and grabbed a musket and charged up the ridge with the Regt. and was one of the very first men in the Brigade that reached the top of the ridge; that just as said King reached the top he was wounded by a musket ball in [the] right arm, breaking the bone. Deponent further says that he was in active service for four years and while in such service witnessed many acts of bravery on the part of soldiers and that this brave charge of said King, outside of the line of his duty and knowing the great danger he was facing, was one of the bravest acts that fell under notice of deponent.

Simeon D. Long, being duly sworn, deposes and says: I was a member of Co. D, 11th Mich. Inf. and served in the War of the Rebellion, with James W. King, Q.M. Sergeant of that regiment, from Aug. 24, 1861, to Sept. 30, 1864. As Q.M. Sergeant his duties were with the stores. Before he could go into battle with his comrades, permission must be obtained from the commanding officer. The night before the charge of Mission Ridge he obtained permission from Gen. [then Colonel] Stoughton, in command of [the] brigade, to take part in the expected battle on the following day. I know from my own personal knowledge that this action was voluntary on Comrade King's part, and that not only Gen. Stoughton, but Major Bennett, who had command of the regiment, told him that 'it was not in the line of his duty,' and tried to dissuade him from his perilous undertaking. On the 25th of November, 1863, Comrade King, armed with a Springfield rifle, remained with the regiment in line of battle from early morn until 4 o'clock P.M. when the order was given for the Army of the Cumberland to charge. His command crossed an open field for half a mile, swept with musket balls and shell, and the said King was among the first to leap the Confederate earthworks at the foot of the ridge. Our command remained in the works for two or three minutes and then started for the crest of the mountain, three quarters of a mile away, at an angle of more than 45 degrees. The top of the ridge was a blaze of fire from artillery and musketry, and showers of canister and musket balls swept the steep ascent. Comrade King was one of the leaders through this storm of death and destruction, and was among the first to reach the mountain summit. Detachments of the 11th Mich. Inf., 19th Illinois, and 69th Ohio, broke the Confederate line in our front. Just to our right the Confederate line was still intact, and not more than five or six rods away was a semi-circular earthwork where a battery of six guns was being worked on the Union troops farther down the ridge. There was also a support of 400 or 500 Confederate infantry, who were keeping up a murderous fire on our men below. There was desperate fighting at this point for the possession of these guns in which Comrade King took a gallant part, and, just as the Union troops were victorious, a Minie ball broke and crushed the bone of his right arm above the elbow joint. He could do no more fighting on his own account and, after having a handkerchief bound tightly around his arm, while the rest of the comrades held the ground so stubbornly fought for and so dearly gained, with right arm mangled and dangling by his side, he passed down the ridge about 15 rods, to the main body of the regiment, to have Major Bennett hurry the rest of the men to the top of the mountain. The major had been killed and Capt. P. H. Keegan was in command. Comrade King explained the situation, that the Confederates had deserted their guns, but that they might return and then we would need help to hold them. Comrade King returned to the top of the mountain the second time where he remained until the routed and flying Confederates disappeared from view down the eastern slope of the ridge.

It is a well-known fact among the Comrades of the regiment that Comrade King was on the mountain top until between sun-down and dark, when with his Springfield rifle on his left shoulder, he passed down the cragged side of the mountain, walked three miles to our camp on Cameron hill, where he left his gun and accoutrements, and then walked a distance of a mile to general hospital. There was a council of surgeons and all said his right arm must come off except our regimental surgeon, Dr. Elliott, and through his intercession it was saved.

I participated in all the desperate battles in which the regiment was engaged, among them Stone River, Chickamauga, and Mission Ridge, and witnessed for over three years the brave acts of the bravest soldiers that ever fought on [a] battlefield, and never, during all that time, did I witness a braver act or a more heroic deed than was performed by Comrade King in the voluntary part taken by him in the charge of Mission Ridge.

James W. King's
Revolutionary War Ancestry

James King was asked for autobiographical sketches on several occasions later in life, and he invariably inaugurated his life story by mentioning the Revolutionary War service of his great-grandfather and role model, Benjamin Montanye.

On March 29, 1781, Montanye, a Continental Army post rider, guided his horse through Ramapo Pass near the New York–New Jersey border. The messenger, under direct orders from George Washington, was entrusted with a large cache of letters, including many bearing the general's signature.[1]

A band of Tories waylaid Montanye. They shot the horse's leg, seized its rider, and dragged his doomed mount off into the woods. The assailants were led by James Moody, a troublesome loyalist later referred to by General Washington as "that Villain Moody." Montanye and his captors endured a debilitating journey through forest and swampland in bitter winter weather and with inadequate supplies, but three days later they arrived in New York City. "Another Rebel mail has been intercepted," declared the British commander in chief Sir Henry Clinton, "which has given me very important information." Moody was promoted to lieutenant and shared with his companions a 100-guinea reward from the British. Montanye, on the other hand, suffered imprisonment in one of the infamous sugar house prisons in New York City. The April 4, 1781, issue of the Tory *New York Gazette* carried the news of his capture along with the text of one of Washington's intercepted letters, and London's *Political Magazine* deemed it all worthy of reprint two months later.[2]

Yet Montanye may not have suffered in vain. The post rider later affirmed that Washington personally directed him to follow a route upon which he was

King-Babcock Family Tree

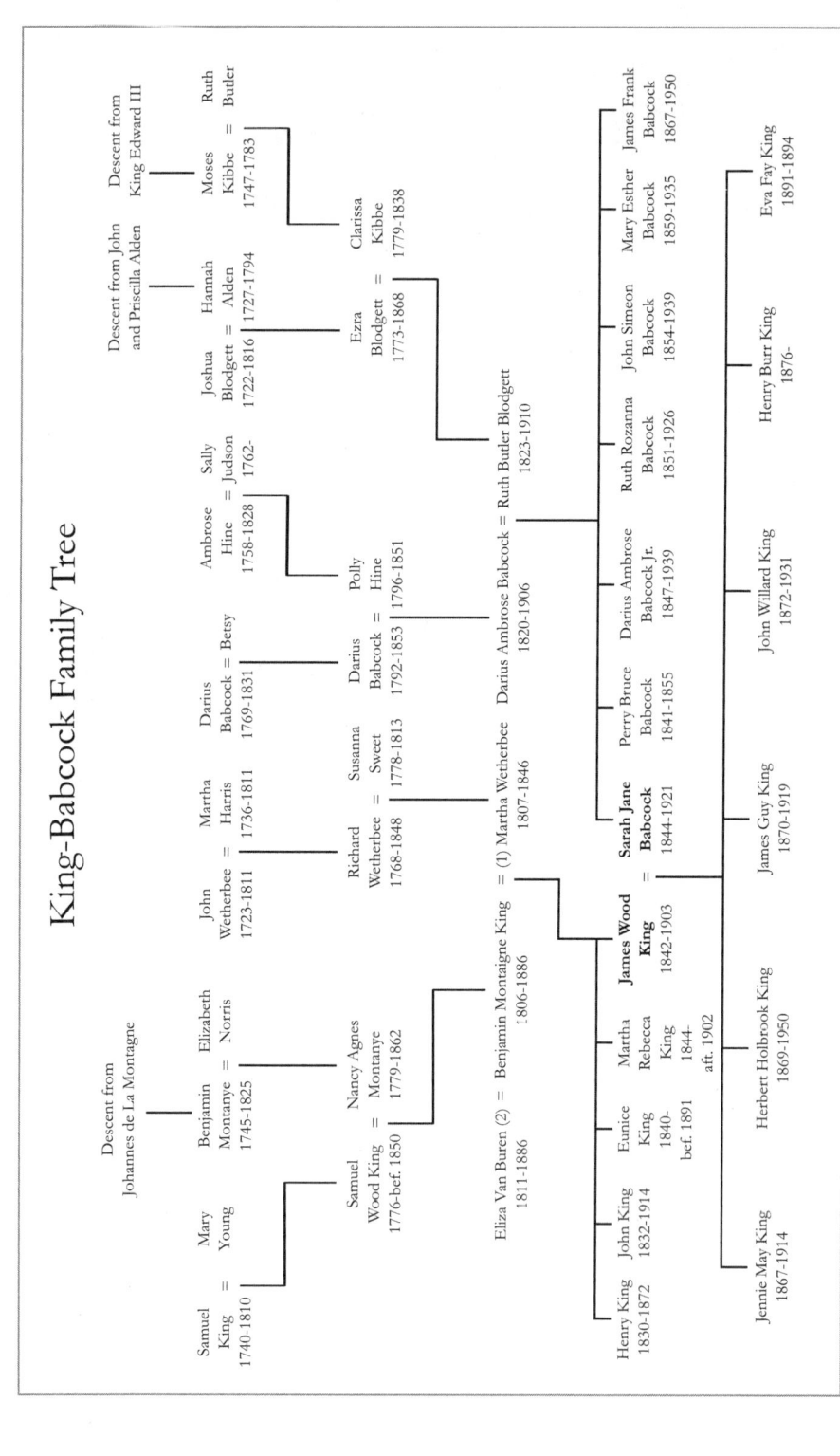

sure to be captured—indeed, the pass was known to be infested with **Tories**—implying that the intercepted correspondence was actually intended for British eyes. Some historians have since suggested or asserted that Montanye was part of a ploy Washington described after the war.[3]

> It was determined by me (nearly twelve months beforehand) at all hazards to give out and cause it to be believed by the highest military as well as civil officers that New York was the destined place of attack, for the important purpose of inducing the eastern and middle States to make greater exertions in furnishing specific supplies than they otherwise would have done, as well as for the interesting purpose of rendering the enemy less prepared elsewhere. . . .
>
> That much trouble was taken and finesse used to misguide and bewilder Sir Henry Clinton in regard to the real object, by fictitious communications . . . is certain, nor were less pains taken to deceive our own army; for I had always conceived, when the imposition did not completely take place at home, it could never sufficiently succeed abroad.[4]

Among the letters captured with Montanye was one in which Washington wrote to inform Benjamin Harrison of Virginia that it was unrealistic for additional reinforcements to be sent south, and that "the most powerful diversion that can be made in favor of the Southern states, will be a respectable force in the neighborhood of New York." Clinton, after scrutinizing this and additional correspondence seized by Moody in May, responded by writing from New York on June 8, "From all the letters I have seen, I am of [the] opinion . . . that the enemy will *certainly* attack this post." But that August, the Continentals would slip away from their unsuspecting enemy in New York and march south to their decisive victory at Yorktown, Virginia.[5]

The post rider was eventually freed in a prisoner exchange. He survived the war to raise a family, and to found and preach at a Baptist church in New Vernon, New York. Montanye's daughter Nancy married Samuel King, a fellow New Yorker born shortly before the Declaration of Independence was signed. Samuel was the son of another Samuel King who served as a lieutenant in the New York militia during the Revolution.[6]

Nancy and Samuel King named their son born in 1806 after his heroic grandfather. Benjamin Montaigne King—James W. King's father—grew up hearing tales of George Washington and James Moody while seated in his grandfather's lap.[7]

Notes

Introduction

1. For more about the motivations that drove Union soldiers to enlist, see McPherson, *For Cause and Comrades*; Wiley, *Life of Billy Yank*; Mitchell, *Civil War Soldiers*; and Gallagher, *The Union War*. For a counterargument asserting the importance of slavery as a motivator, see Manning, *What This Cruel War Was Over*.

2. The rise of this stereotype, and the rewriting of Reconstruction history in general, is described in Foner, *Reconstruction*, 294–96, 608–10.

3. Van Buren was the brother of James's stepmother, Eliza. In addition to Van Buren's book (*Jottings of a Year's Sojourn in the South, or, First Impressions of the Country and Its People; With a Glimpse at School-Teaching in That Southern Land, and Reminiscences of Distinguished Men*), he authored scores of articles for Michigan's pioneer society, covering such topics as education, politics, and temperance. Van Buren's correspondence (including James King's letter of March 30, 1863, appearing later in this book) is preserved at University of Michigan's Bentley Historical Library.

4. For more about the history of courtship in America and on the related themes discussed herein, see Rothman, *Hands and Hearts*. For the growing acceptance of romantic love at mid-century, see pp. 103–4 in that work.

5. The details of everyday life for a Union soldier are covered extensively in Wiley, *Life of Billy Yank*.

6. For more on the wartime evolution of Northern opinions on race, see McPherson, "Slavery Must Be Cleaned Out," chap. 9 in *For Cause and Comrades*; and Mitchell, *Civil War Soldiers*, 117–30.

7. Michigan's role in the war is the subject of Richard Bak's *A Distant Thunder* and Jack Dempsey's *Michigan and the Civil War*. The thousand-page opus *Michigan in the War*, compiled by John Robertson and distributed free of charge to the state's

Civil War veterans, remains a treasured source. Published Michigan Civil War letters, diaries, and memoirs exist, but in relatively limited numbers. Some examples are Ely, *Diary of Captain Ralph Ely of the Eighth Michigan Infantry*; Edmonds, *Memoirs of a Soldier, Nurse, and Spy*; Kimball, *Among the Enemy*; Mayo, *Civil War Letters of Perry Mayo*; Paddington, *Dear Sarah*; Sears, *For Country, Cause and Leader*; Willcox, *Forgotten Valor*; Wilterdink, *My Country and Cross*; and Wittenberg, *At Custer's Side.*

8. The best published sources for the 11th Michigan Infantry are Thornton, *When Gallantry Was Commonplace*, and *Diary of Ira Gillaspie*. Thornton's regimental history is a solid effort but has become dated in its battle narratives, due to the wealth of scholarship published on the Western Theater in recent years. Gillaspie's diary represents an excellent primary source but lacks annotations, offers only a cursory introduction, and stops early in 1863. An excellent master's thesis on the regiment, "The Road to Murfreesboro," was authored by Wayne C. Mann at Western Michigan University in 1963, but that work covers only the period from the regiment's organization through the Battle of Stones River.

9. For more about the postwar lives of Civil War veterans, including discussion of the challenges faced by disabled veterans, see Marten, *Sing Not War*; Logue and Barton, *The Civil War Veteran*; and Nelson, "Empty Sleeves and Government Legs: The Ruins of Men," chap. 4 in *Ruin Nation*.

1. Drum and Fife

1. Benjamin King served in Captain Henry Powers's company from May 21 through June 21, 1832. The 120 acres he purchased is located in present-day Three Rivers, on the southwest corner of Broadway Road and U.S. Highway 131. Silliman, *St. Joseph in Homespun*, 129–30; Van Buren, "Deacon Isaac Mason's Recollections," 5:399–400; Barnett and Rosentreter, *Michigan's Early Military Forces*, 153, 197.

2. Morrison, *King Genealogy*, 7; Silliman, *St. Joseph in Homespun*, 130; Riker, *Revised History of Harlem*, 784–90.

3. Benjamin and Martha (Wetherbee) King had eight children in all, but three of them died young, and their names are unrecorded. James was their sixth offspring. Eliza's family was among the earliest settlers of Battle Creek. Her mother, Olive Jay, was descended from Founding Father John Jay. King, "Another Pioneer Gone," 10:184; *Portrait and Biographical Record of Kalamazoo*, 645–46; *American Biographical History*, 62.

4. This house stood for 150 years before suffering demolition in favor of a hardware store. King, "Another Pioneer Gone," 10:184; Silliman, *St. Joseph in Homespun*, 129; James W. King, "Stranger Than Romance," n.d., p. 1, James W. King Collection, 1861–1903, Western Michigan University Archives and Regional History Collections (hereafter cited as King Collection).

5. "James W. King," 2; *Portrait and Biographical Album*, 267; "Opinions of James W. King," *Three Rivers (Mich.) News Reporter* (hereafter cited as *News Reporter*), Oct. 15, 1903, photocopy in the King Collection.

6. The representatives were Joshua Blodgett and Captain Ambrose Hine. Babcock, *Babcock Genealogy*, 1, 93–94; Application for Membership, Daughters of the American Revolution, National Nos. 392452 and 481789, DAR Library, Washington, D.C.;

Harrison, *Family Forest Descendants of Thomas Blodgett*, 4, 52; Hanna, *Kibbe Genealogical Notes*, 59–60; Lillian S. Dudd, Babcock Genealogy, John Dudd Collection, 1958, Western Michigan University Archives and Regional History Collections (hereafter cited as Dudd Collection); Johnston, *Record of Connecticut Men*, 231, 406, 522, 624; *Roll of State Officers*, 26–29.

7. Clarissa's parents were Moses and Ruth (Butler) Kibbe. The royal lineage, traced through Moses, is well documented and includes scores of medieval monarchs. Harrison, *Family Forest Descendants of John Alden*, 293; Hanna, *Kibbe Genealogical Notes*, 59–60; Harrison, *Family Forest Descendants of Edward III*, 2884; Roberts, *Royal Descents*, 1:149, 152.

8. For more on the initial Northern reaction to Fort Sumter, see Wiley, *Life of Billy Yank*, 17–21. For more about Montanye, see appendix B. King, "Another Pioneer Gone," 10:184; "The War Feeling in Three Rivers," *Three Rivers (Mich.) Western Chronicle* (hereafter cited as *Chronicle*), Apr. 25, 1861; *Portrait and Biographical Album*, 266; Application for Membership, Daughters of the American Revolution, National No. 141450, DAR Library, Washington, D.C.; McPherson, *For Cause and Comrades*, 18–19, 104–5.

9. In the wake of the Union defeat at Bull Run, the federal government, desperate to raise additional troops quickly, perceived the state governments as unable to comply fast enough. In response, commissions were granted for individuals to recruit units independent of their state governments. This experiment was short lived; the independent units were turned over to state control in September. *War of the Rebellion*, series 3, vol. 1, pt. 1, pp. 518–19 (hereafter cited as *OR*).

10. Camp Tilden was located just south of town at facilities donated for military use by the Michigan Southern Railroad. A single car-house, six hundred feet long, served as the regimental quarters. "Col. May's St. Jo. Regiment," *Chronicle*, Sept. 11, 1861.

11. In volunteer units it was customary for regimental officers to be elected by their men, and for all ranks above lieutenant to be voted to the individuals most active in recruiting the regiment. May had served as colonel of two militia companies and was a major player in raising the regiment. Stoughton brought some modest martial experience to the table as well, having captained the Sturgis Peninsular Guard. He had attended the National Republican Convention in 1860 and was a firm supporter of the party. He resigned his position as United States district attorney for Michigan—an appointment bestowed by Abraham Lincoln—in order to serve his country. Mann, "Road to Murfreesboro," 22; "St. Jo. Independent Regiment," *Constantine (Mich.) Weekly Mercury* (hereafter cited as *Mercury*), Sept. 19, 1861; Thornton, *When Gallantry Was Commonplace*, 19; "Peninsular Guards," *Chronicle*, May 2, 1861; Fitch, *Army of the Cumberland*, 238.

12. Odd was a nickname, probably for Private William H. Davis of Company A. Later letters demonstrate that Davis was a friend of the Babcock family.

13. Captain Henry Spencer's Three Rivers Light Guard, marching to the tunes of the Three Rivers Brass Band, was seen off by a large crowd of admirers at the railroad depot. The guard was later designated Company E. In the early stages of recruitment, each company sported its own name; upon formal organization of the regiment, these

names were replaced with letter designations. "Departure of the Three Rivers Light Guard," *Chronicle*, Sept. 11, 1861; "Col. May's St. Jo. Regiment," *Chronicle*, Sept. 11, 1861; "Our Army Correspondence," *Chronicle*, Nov. 13, 1861.

14. Colonel May was ordered to give up three hundred of his men to fill the ranks of another regiment. He protested, and the number was reduced to eighty-three. "Col. May's Independent Regiment," *Chronicle*, Sept. 18, 1861.

15. Jenny's first cousin, Corporal Lemuel Packard Pierce of Fabius, was the son of David and Rheuma (Blodgett) Pierce. He was another of Ezra Blodgett's eighty or so grandchildren. He mustered into Company A with James. *Record of Service of Michigan Volunteers*, 11:73.

16. For more about a typical day's routine for a regiment far removed from the fighting, see Wiley, *Life of Billy Yank*, 45–47.

17. The identification of W. S. Wood is problematic, but the most likely candidate appears to be Corporal William Woods of Wayne County, Michigan, who enlisted in Company I of the 4th Michigan Infantry on June 20, 1861. The 4th Michigan was stationed at Washington, D.C., in this period. *Record of Service of Michigan Volunteers*, 4:125.

18. Captain David Oakes Jr. of Company A.

19. Austin Blair was a founding member of the Republican Party and a staunch Unionist. He had pushed for black suffrage in Michigan before the war and was a steady opponent of slavery. Despite a failed bid for the U.S. Senate in 1858, Blair was elected governor of Michigan in 1860. He worked tirelessly to meet his state's recruiting goals throughout the war; the 1st Michigan Infantry arrived in Washington, D.C., one month after Lincoln's first call for troops.

20. Sergeant Cuthbert Dixon of Company A went by the nicknames Dick, Dix, and Dixie. He appears in the 1860 census in Constantine, Michigan, an eighteen-year-old native of England living with a family by the name of Hartman. The patriarch of the family, Philip, was from Pennsylvania, but the woman of the house, Elizabeth Hartman, was also born in England. Possibly she was Dixon's mother and had remarried. *Record of Service of Michigan Volunteers*, 11:29; 1860 United States Census, Schedule 1, Constantine Township, St. Joseph County, Mich., dwelling 782, family 790, July 10, 1860.

21. Gambling and profanity were unavoidable aspects of soldiering. The latter was explicitly forbidden by the articles of war, but enforcement was a different matter entirely. For more, see Wiley, *Life of Billy Yank*, 248–52; and Lorien Foote, "'A Good Moral Regiment': Conduct Unbecoming a Gentleman," chap. 1 in *The Gentlemen and the Roughs*.

22. Charles P. Buck would enlist as a private in the 25th Michigan Infantry in August 1862, and muster out in June 1865 as a sergeant. He was a son of Quakers George and Martha (Irey) Buck. George Buck, born in Pennsylvania in 1792, platted the village of St. Joseph (later part of Three Rivers) in 1830 along with Jacob McInterfer, whose house the King family rented in 1834. The Bucks operated the first tavern in town, and George held a stake in a local bank in addition to other business interests. James King's father, Benjamin, once related a tale of the Panic of 1837 in which "Mr. Buck was notified that the bank examiners were on their way to the village. He went to Kalamazoo to borrow gold and silver deposits to be ready for them. He went with a one horse

wagon and on returning with the kegs of deposits lost his way and had to remain in the forest overnight. He slept in his wagon with the gold for a pillow but reached the bank ahead of the examiners." George's wife, Martha, born near Harper's Ferry in 1798, was a first cousin of the radical abolitionist John Brown. *Record of Service of Michigan Volunteers*, 25:14; 1850 United States Census, Schedule 1, Lockport, St. Joseph County, Mich., dwelling 54, family 54, Aug. 29, 1850; Silliman, *St. Joseph in Homespun*, 53–56.

23. The visit by Michigan governor Austin Blair was eagerly anticipated for October 1, and although he failed to show that day, optimism prevailed, and the lone cannon in camp was polished to a shine two days later in anticipation of the rescheduled event. The rank and file dressed up such as they could without uniforms, and the officers mounted their horses. The train again arrived sans politician, and a rainstorm shooed the disappointed troops indoors. Blair finally showed up on the eighth of the month. Stoughton drilled the soldiers, and May reviewed them with the governor, who proclaimed the men of the 11th Michigan among the best the state could offer. "Col. May's Independent Regiment," *Chronicle*, Oct. 9, 1861; "The Governor's Visit," *Reporter*, Oct. 12, 1861; Mann, "Road to Murfreesboro," 25–26.

24. Miss Sutton was a former teacher at the district school. The 1860 census shows an Elizabeth Sutton living in Lockport. Susan M. Wells to Benjamin F. Wells, Jan. 25, 1862, Wells Family Papers, 1857–1902, Bentley Historical Library (hereafter cited as Wells Papers); Sarah J. Babcock to James W. King, Mar. 23, 1862, King Collection; 1860 United States Census, Schedule 1, Lockport Township, St. Joseph County, Mich., dwelling 1474, family 1490, Aug. 6, 1860.

25. Company clerks, detailed from the rank and file, were responsible for maintaining books, reports, returns, rolls, requisitions, and other paperwork for their captain. Ideally, a clerk would be selected for his superior organizational skills and penmanship. The perks of the job included being excused from some of the more tedious aspects of soldiering, such as guard and work details. Kautz, *The Company Clerk*, 9–10.

26. Bounties of land and money were offered as enlistment bonuses. Private Ira Gillaspie of Company C noted in his diary that his recruiting officer touted the enticements of 160 acres of land and $100. Although soldiers' pay was modest, the economy was in a downturn, and some men did enlist for the money. *Diary of Ira Gillaspie*, 7; Wiley, *Life of Billy Yank*, 37–38.

27. It was common for a soldier to face the dilemma of his parent, wife, or sweetheart begging him to finagle a discharge, and James's response is typical of the men who faced such appeals. Most Union soldiers believed they were defending the very existence of the United States as they knew it; there was little sense in returning home a coward when country, family, and friends would all be in peril without men willing to fight. Like many of his peers, James additionally invoked the concept of duty. Duty was a sacred word in Victorian America, and Union soldiers relied on its magic to persuade the folks at home that they were doing the right thing by serving their country. McPherson, *For Cause and Comrades*, 22–23, 99. Note that McPherson quotes from this letter of October 18.

28. Billy could not be identified. He evidently enlisted between October 18 and James's next mention of him on November 6, but efforts to identify him on the

regimental roster have failed. James refers to other comrades named Billy in later correspondence, but none of those individuals are a sure match.

29. A portion of the needed uniforms had been shipped by the War Department, but they were snagged in a glut of supplies piling up in Pittsburgh. The government was overwhelmed by the task of clothing and equipping an army, and such delays were only typical. *Reporter*, Oct. 26, 1861, p. 3.

30. The regiment had received its official designation on October 11. "May's Regiment the Eleventh," *Chronicle*, Oct. 16, 1861.

31. George F. Grather of Centreville was a private in Company A. He lost a battle with typhoid on October 19. His was the 11th's first death. Regimental Service Records, 11th Michigan Infantry, Archives of Michigan (hereafter cited as Regimental Service Records).

32. Roughly quoting the poem "Rodieker's Youth" by Walter Colton.

33. Most of the uniforms still had not arrived. A Michigan company was asked to fill the order after it was discovered that a Philadelphia government contractor shared responsibility for the delay. "The old R.R. Dining House," one reporter remarked, "presented indeed a lively appearance with its swarms of soldiers partly uniformed.'" "Col. May's Independent Regiment," *Chronicle*, Nov. 27, 1861; "Villages and Improvements," *Chronicle*, Nov. 13, 1861.

34. Daniel Webster Holbrook of Company A was a resident of Branch County. His parents, Silas Arms and Emelia (Hickok) Holbrook, were among the early settlers of Coldwater, Michigan, having immigrated in 1831. Holbrook's early childhood memories included climbing out of bed one morning and stepping on a sleeping Indian who had snuck into the cabin on a cold night, seeking shelter. Silas Holbrook held a number of prestigious positions in town, including associate judge, postmaster, and village president. He went west in 1851 for the gold rush and died in California the following year. *Record of Service of Michigan Volunteers*, 11:49; *History of Branch County*, 240–41.

35. A false rumor.

36. William Clarence Iddings, a Company A private, was a Pennsylvania native. *Record of Service of Michigan Volunteers*, 11:51.

37. Benjamin Franklin Wells of Fabius was another member of Company A. He was born into poverty in Flat Rock, Ohio. Wells's son later wrote of his father's childhood, "His first pair of shoes was earned by gathering hickory nuts. . . . The first home Benjamin's parents ever owned was the fruit of his toil while he was still a mere boy. . . . Every privilege he ever had was of his own earning." In his mid-teens, the affable Wells immigrated with an older boy to Three Rivers, where he apprenticed as a mason and proved capable in his trade. He persevered to work his way through high school and became a district schoolteacher before the war. *Record of Service of Michigan Volunteers*, 11:100; Wells, *My Father Benjamin F. Wells*, 5–7.

38. First Sergeant William W. Hoisington of Company E, also known as Wallace or Walter, was both scholarly and "the finest specimen of physical manhood" in St. Joseph County, as Benjamin Wells's son later wrote. "None could lay him on his back in a friendly bout." Wells married Hoisington's sister, Susan Melissa Hoisington, on August 27, just before going off to war. Hoisington's fifty-eight-year-old father Abishai,

considered one of the best snare drummers in the county, enlisted as a drummer in Company G and was later appointed drum major. Wells, *My Father Benjamin F. Wells*, 8; *History of St. Joseph County*, 212–13.

39. Loriston Alden Fulkerson of Three Rivers, a Company A private, was Jenny and Lemuel Pierce's first cousin. He was a son of Reverend Edward and Hannah (Blodgett) Fulkerson. The name Alden has been passed down among the descendants of Mayflower Pilgrims John and Priscilla Alden. *Record of Service of Michigan Volunteers*, 11:38.

40. Smallpox vaccination was mandatory under army regulations.

41. James W. King, "Reunion of Michigan Veterans," unidentified newspaper, Aug. 25, 1873, photocopy in the King Collection; "Col. May's Independent Regiment," *Chronicle*, Dec. 4, 1861; "Eleventh Regt. Mich. Infantry," *Reporter*, Nov. 30, 1861.

42. Colonel May requested that the regiment be deployed to South Carolina instead, in hopes of seeing action as soon as possible, but without success. As demonstrated in this letter, it was commonplace for false rumors of a unit's deployment to precede its actual exit from camp. "Col. May's Regiment," *Chronicle*, Nov. 27, 1861; Wiley, *Life of Billy Yank*, 30.

43. James refers to First Sergeant Aaron B. Sturges of Company A. Drunkenness was a major disciplinary issue in the Union army. See Wiley, *Life of Billy Yank*, 252–54; Foote, "'A Good Moral Regiment'"; and Ramold, "'Brawls, Riots, and Midnight Orgies': Alcohol and the Union Army," chap. 4 in *Baring the Iron Hand. Record of Service of Michigan Volunteers*, 11:90.

44. The sutler was Marshall M. Wells, who had served as a representative for Oakland County, and as deputy U.S. marshal under President Buchanan. The wounded soldier, Private Charles Leonard of Company C, recovered and stayed with the regiment. *Diary of Ira Gillaspie*, 12; Bingham, *Early History of Michigan*, 681; *Record of Service of Michigan Volunteers*, 11:58.

45. "From the 11th Michigan Regiment," *Chronicle*, Dec. 25, 1861; James W. King, "11th Michigan Infantry," *Lansing (Mich.) State Republican* (hereafter cited as *State Republican*), Jan. 11, 1901, photocopy in the King Collection; *Annual Report Adjutant General 1862*, 37; James W. King, "Reunion of Michigan Veterans," unidentified newspaper, Aug. 25, 1873; *Diary of Ira Gillaspie*, 12–13; James W. King to Sarah J. Babcock, Apr. 16, 1865, King Collection.

46. Don Carlos Buell, West Point class of 1841, was a hero of the Mexican War. A brave soldier and capable strategist, he showed promise in the early months of the Civil War. He was commissioned brigadier early in the conflict, was granted division command in August 1861, and succeeded Sherman to command the Department of the Ohio in November 1861. For more, see Engle, *Don Carlos Buell*. Kentuckian Simon Bolivar Buckner, West Point class of 1844, served as an instructor at the military academy and fought with distinction in the Mexican War. In 1861 he championed the preservation of Kentucky's neutrality until troops from both sides entered the state, at which time Buckner reported to General Albert Sidney Johnston at Bowling Green, and was commissioned brigadier general.

47. Martha Rebecca King, James's younger sister, was born in October 1844. She was still living with their parents at this time.

48. Not everyone celebrated the bluecoats' arrival. "There are about four thousand inhabitants in this town," said Charles Rice, "and one-half of them are 'secesh' but they have to keep mum." "From the 11th Regimental Band," *Chronicle*, Jan. 15, 1862.

49. The daily ration at this time consisted of pork or beef, flour, hardtack, vegetables, coffee, sugar, salt, and vinegar. The ration was impressive by contemporary military standards, but quality varied, the nutritional balance was poor, and supply issues often rendered the prescribed list of victuals a fantasy. Adams, *Doctors in Blue*, 206–9.

50. Paraphrasing Martin Farquhar Tupper's "Of Cheerfulness."

51. Charles Rice reported that the regimental band brightened Christmas Eve with a "Grand Serenade." The soldiers visited some of the more notable local citizens, and Rice declared that "they had everything to eat and drink that heart could wish. I never saw people with more cordiality and friendship than they do in Kentucky." Some of the soldiers took up a collection to buy turkeys for Christmas dinner, and half a dozen blacks staged a musical performance featuring banjos, tambourines, and fiddles. The Christmas night festivities were later topped off with a display of fireworks. "From the 11th Michigan Regiment," *Chronicle*, Mar. 5, 1862; *Diary of Ira Gillaspie*, 14.

52. The Sanitary Commission recommended that each company hire a professional cook, but in practice the food was prepared by amateurs who were pulled from the ranks and did little more than prepare coffee and beans. Adams, *Doctors in Blue*, 211.

53. James's brother John had moved to Waukegan, Illinois, in 1858. James W. King, "Truth Is Stranger Than Fiction," p. 1, King Collection.

54. Brigadier General Thomas John Wood assumed the role of camp commander on December 24. He relocated the 11th to join the other soldiers under his command and took measures to instill discipline into the green troops. The men were required to assemble at reveille for roll call and strict inspection, and regimental commanders were held accountable for ensuring that their soldiers were properly equipped and supplied with ammunition. Thornton, *When Gallantry Was Commonplace*, 65–66.

2. RATIONS AND COFFINS

1. For more about how Americans coped with death as a constant companion during the war, see Faust, *This Republic of Suffering*. For discussion of rural units and disease, reference Wiley, *Life of Billy Yank*, 124–28, 132–33. James W. King, "Reunion of Michigan Veterans," unidentified newspaper, Aug. 25, 1873.

2. The rumor mill in Three Rivers would falsely report Second Lieutenant Henry S. Fisher dead later in the month. A number of the men had come down with smallpox, measles, and respiratory illnesses in recent days. Although smallpox vaccination was mandatory, compliance among the undisciplined volunteer regiments varied, and the 11th would suffer an abnormally high number of cases compared with the Union army overall. *Record of Service of Michigan Volunteers*, 11:36; "Head Quarters 11th Mich. Inf.," *Reporter*, Feb. 1, 1862; *Diary of Ira Gillaspie*, 12–13, 15; Thornton, *When Gallantry Was Commonplace*, 67.

3. The uncle in question is probably Anson De Puy Van Buren. Later letters show that he frequented Fabius.

4. Closely following a passage from "The Lady of the Lake" by Sir Walter Scott.

5. Ephraim Elmer Ellsworth, an associate and friend of Abraham Lincoln, directed a famous nationwide tour of Zouave drill performances before the war. Ellsworth was shot dead by James W. Jackson on May 24, 1861, in retaliation for taking down a Confederate flag mounted on the roof of Jackson's tavern in Alexandria, Virginia. James's statement regarding Jackson's brother appears correct: The *Constantine Mercury* quoted the *Bardstown Gazette* stating that the 11th's camp was on the grounds of one Dr. Jackson. James Jackson's brother, Dr. John Jackson, lived in the area and was arrested in November 1861 for shooting a Union soldier. "From the Eleventh Regiment," *Mercury*, Mar. 13, 1862; *Life of James W. Jackson*, 13.

6. From Tennyson's "The Grandmother."

7. The unit had received additional exposure to measles while relocating camp. A group of sick Indiana soldiers, some carrying the dreaded disease, passed through the 11th's ranks during its march. Corporal James Martin of Company H wryly observed, "There wasent small pox enough to go around, so we have got to take up with the measles.... It is a hard place to be sick in camp a sick man isent no more cared for than a sick calf." As bad as things were, reports of disease fatalities were exaggerated in the newspapers back home. In response, Captain Benjamin Grove Bennett of Company D began one letter to the editor of the *Western Chronicle*, "From this you will see that there is at least one man yet alive in the Michigan 11th." "From the Eleventh Regiment," *Mercury*, Mar. 13, 1862; Benjamin F. Bordner to his brother, Jan. 15, 1862, Benjamin F. Bordner Papers, 1862–1864, Bentley Historical Library (hereafter cited as Bordner Papers); James Martin to his parents, Jan. 11, 1862, James Martin Letters, 1861–1864, Bentley Historical Library (hereafter cited as Martin Letters); "From the 11th Michigan Regiment," *Chronicle*, Feb. 5, 1862.

8. James Abram Garfield, future twentieth president of the United States, was a Republican senator in the Ohio legislature when the war started. He secured the colonelcy of the 42nd Ohio and was granted brigade command by Don Carlos Buell. Garfield defeated Brigadier General Humphrey Marshall at Middle Creek, Kentucky, on January 10 and was promptly promoted to brigadier general. For more about Garfield, see Millard, *Destiny of the Republic*.

9. The regimental hospital in this period was treating two hundred patients at times. Measles was the biggest killer. "Army Intelligence," *Reporter*, Feb. 15, 1862.

10. Confederate brigadier general Felix Kirk Zollicoffer, former Tennessee state senator and U.S. congressman, was slain when he mistakenly advanced into a Union position during a bungled attack against George H. Thomas at Mill Springs, Kentucky, on January 19.

11. Alden and Lewis had already lost their mother, Hannah (Blodgett) Fulkerson, before the war. Their father, Reverend Edward Fulkerson, passed away on January 16. Lewis had enlisted in the 2nd Michigan Infantry when the war broke out but was discharged for disability at Fort Lyon, Virginia, on November 24, 1861. *Record of Service of Michigan Volunteers*, 2:71.

12. This seems to be a combination of two very rough extracts, in reversed order, borrowed from Sir Walter Scott's epic poem *Marmion* and paraphrased by James's uncle, Anson De Puy Van Buren, in his book *Jottings of a Year's Sojourn in the South*.

De Puy's quote departs from Scott's original text, and James further alters that wording slightly in his rendition. The original reads:

A forest glade, which varying still,
Here gave a view of dale and hill,

....

Hill, brook, nor dell, nor rock, nor stone,
Lies in the path to me unknown.

Van Buren, *Jottings*, 95.

13. Haight "was esteemed by the men of the company," Adjutant Samuel Chadwick wrote, "as a kind, efficient officer, and was beloved by every man." Fellow officers and medical staff had implored Haight to take a medical furlough, but the lieutenant declined after his men begged him to stay with the regiment. "Camp Morton, near Bardstown, Ky.," *Reporter*, Feb. 22, 1862; "Death of Lieut. Haight," *Reporter*, Feb. 22, 1862.

14. Captain Oakes described the slave huts of the Bardstown area as "generally log cabins, low and poorly furnished, not as comfortable in many cases, as the pig pens of northern farmers." "Head Quarters 11th Mich. Inf.," *Reporter*, Feb. 1, 1862.

15. Christopher Haight's Pension Application, Application No. 360, Certificate No. 5444, National Archives; James W. King's Compiled Military Service File, National Archives; "From the 11th Michigan Regiment," *Chronicle*, Feb. 26, 1862; "Camp Morton, near Bardstown, Ky.," *Reporter*, Feb. 22, 1862.

16. Regimental records list measles as the cause of Alden's death. Erysipelas was a major killer in the unsanitary Civil War hospitals, with a fatality rate exceeding 40 percent. For more, see Devine, "Civil War Bodies and the Development of Experimental Method: Erysipelas and Hospital Gangrene during the American Civil War," chap. 3 in *Learning from the Wounded*. Regimental Service Records.

17. Before the war, Henry Hopkins Sibley invented the Sibley tent and designed a stove for heating the shelter. Sibley became a Confederate brigadier general who led a failed campaign in New Mexico that culminated in a decisive strategic Union victory at the Battle of Glorieta Pass on March 28, 1862. The 11th Michigan received the spacious, conical Sibley tents upon their arrival in Louisville. The shelters were intended to house about a dozen men but in practice often lodged many more. Hicks, "Personal Recollections," 522–23; Wiley, *Life of Billy Yank*, 55.

18. The 1870 census lists the family of Horatio Unsell in Bardstown. 1870 United States Census, Schedule 1, Bardstown, Nelson County, Ky., p. 9, dwelling 71, family 88, June 2, 1870.

19. Private Daniel Rose of Company A. *Record of Service of Michigan Volunteers*, 11:79.

20. Henry S. Fisher had been promoted to first lieutenant of Company A the same day James was elevated to corporal. Arvin F. Whelan was the regiment's assistant surgeon. *Record of Service of Michigan Volunteers*, 11:36; *History of Hillsdale County*, 118.

21. This was a typical Yankee opinion of life in the South. The Northern view of slavery was rooted in the perception that slaveholders constituted an oligarchy that prospered at the expense not only of the slave but also to the detriment of the common

white Southerner, who was reduced to poverty and ignorance in the absence of a free-labor economy. Mitchell, *Civil War Soldiers*, 109; McPherson, *Cause and Comrades*, 118–19; Wiley, *Life of Billy Yank*, 98–100.

22. Probably referring to Jenny, Lemuel, and Alden's first cousin, Eliza Letitia Dunn. Letitia, daughter of Simeon and Esther (Blodgett) Dunn, is mentioned again later. She married Lemuel Pierce's brother Sylvester after the war.

23. For more about religion and the Civil War soldier, see Woodworth, *While God Is Marching On*. In chapter 10, Woodworth discusses the phenomenon of soldiers turning away from their accustomed civilian religious practices, and associated moral standards, after joining the military.

24. James refers to rumors involving former secretary of war John Buchanan Floyd and Tennessee governor Isham Harris. The claim about Harris—a die-hard Confederate—was false. As for Floyd, he in fact escaped capture, but actual events proved far more dramatic than the rumor. In a matter of days, Grant conquered Confederate forts Henry and Donelson, located on the Tennessee and Cumberland Rivers respectively. At Donelson, Confederate generals Floyd and Gideon Johnson Pillow fled the fort, leaving Simon Bolivar Buckner behind to surrender the army to his prewar friend Grant. Grant famously demanded unconditional surrender, became a household name, and earned promotion to major general. Albert Sidney Johnston was forced to fall back deep into Tennessee. For more, see Gott, *Where the South Lost the War*.

25. Charles Anderson Wickliffe became acting governor of Kentucky upon his predecessor's death in 1839. He represented his state in Congress from 1861 to 1863. His mansion, Wickland, still stands.

26. James W. King, "Reunion of Michigan Veterans," unidentified newspaper, Aug. 25, 1873.

27. On the evening of February 5, Mary Todd Lincoln threw an exorbitant bash at the White House for the Washington elite and the highest echelons of the military. Almost five hundred guests were invited, highly exclusive catering was secured, and wine and champagne flowed freely. The servants were clad in uniforms color-coordinated with Mrs. Lincoln's recently purchased china set. The party went on despite the need for the president and first lady to spend much of the night upstairs at the bedside of their son Willie, who had taken seriously ill. Willie succumbed to his sickness, likely typhoid, fifteen days later. Baker, *Mary Todd Lincoln*, 205–8.

28. James repeats this and a similar saying, "Absence makes the heart grow fonder," in future letters, and the attitude conveyed seems to indicate the key reason his relationship with Jenny thrived through three years of long-distance communication. Between 1861 and 1865, relationships conducted purely via correspondence became a fact of life for millions.

29. A reference to Jenny's younger brother, Darius Ambrose Babcock Jr. The name Darius Babcock was passed down through several generations. Jenny was the eldest of five surviving siblings; the others were Ruth Rozanna, John Simeon, and Mary Esther. The family's firstborn, Perry Bruce, died in an accident in 1855 at age thirteen. Another brother, James Frank, would be born in 1867.

30. Enlisted men were detailed as hospital clerks to ease the administrative burden on dedicated medical personnel. Clerks were excused from their regimental duties.

31. Daguerreian cars were mobile photography studios.

32. The 50th Indiana Infantry was stationed at Bardstown in January and February; presumably the soldier James mentions was left behind sick when the Indianans moved on to Bowling Green in early March. Preparing medical furlough paperwork was one of James's duties as a hospital clerk.

33. Rothman, *Hands and Hearts*, 319n12.

34. Mr. Walton is probably Samuel Walton, an Ohio-born merchant living in the adjacent township of Lockport. 1860 United States Census, Schedule 1, Lockport, St. Joseph County, Mich., dwelling 1491, family 1507, Aug. 6, 1860.

35. Probably Clarinda Fulkerson, Alden's sister.

36. Rebecca King and Letitia Dunn have already been mentioned. Elizabeth cannot be identified with any degree of certainty, but she could be Jenny's first cousin Elizabeth Jane Blodgett, a resident of Fabius and daughter of John Wesley Blodgett. Samuel White was an immediate neighbor of Jenny's, and the 1860 census shows New York native Emma Vanderbilt in his household. Census data suggest that Henrietta was Emma's older sister, who was still living in New York as of 1860. 1860 United States Census, Schedule 1, Fabius, St. Joseph County, Mich., dwelling 1067, 20 July 1860; 1860 United States Census, Schedule 1, Castleton, Richmond County, N.Y., dwelling 625, family 667, June 16, 1860.

37. Doctor Byran could not be identified.

38. George Washington's great-grandnephew, Lieutenant Colonel John Augustine Washington, inherited Mount Vernon and sold it out of the family before the war. He served as aide-de-camp to Robert E. Lee in western Virginia. The Washingtons were in no danger of extinction, but John Washington was killed on September 13, 1861, at Cheat Mountain, shot three times through the torso under the circumstances described in this letter. Lee requested and received the return of Washington's body from the Federals the next day. Joe and Jake were probably members of the 17th Indiana on picket duty at the time of the incident and were now detailed on hospital duty. Keifer, *Slavery and War*, 1:222–25.

39. The rumor about brigading was false. Colonel May resigned effective April 1 and was soon replaced with Lieutenant Colonel Stoughton. May had been in a poor state of health for some time, and he seems to have had some bitter opponents back home, if not within the ranks. The *Three Rivers Reporter* published a false claim on February 22 from "a reliable gentleman now at the camp of the Eleventh Regiment" stating without explanation that May was to be court-martialed. Captain Bennett replied via the *Western Chronicle* shortly thereafter, calling that story "a base falsehood" and suggesting that it was the work of "Col. May's enemys at home." William J. May to Austin Blair, Mar. 29, 1862, in Regimental Service Records; "Col. May Arrested!" *Reporter*, Feb. 22, 1862; "From the 11th Michigan Regiment," *Chronicle*, Mar. 12, 1862.

40. Quoting from "Summer" by James Thomson.

41. A nightmarish bloodletting had taken place near Shiloh Church on April 6 and 7. General Buell marched from Nashville to combine with Grant, who pressed ahead south along the Tennessee River. Confederate forces under Albert Sidney Johnston converged to strike Grant before Buell's arrival. A brutal, chaotic struggle ensued between the inexperienced armies. Grant's men were driven back the first day but counterattacked

the next morning in conjunction with Buell's troops and won a strategic victory. The cost on both sides was staggering; in all, twenty thousand men were killed or wounded. Johnston was among the dead. For more, see Groom, *Shiloh, 1862*.

42. Island No. 10 was a formidable Confederate Mississippi River stronghold located near the New Madrid Bend of the waterway, barring Federal navigation beyond that point. Union general John Pope cooperated with Andrew Foote's gunboats in a well-orchestrated campaign to bag the Rebel garrison. The Confederates surrendered on April 8, the day after the Union counterattack at Shiloh. For more information, see Daniel and Bock, *Island No. 10*.

43. Union major general George Brinton McClellan had launched his Peninsula Campaign, landing the Army of the Potomac in Virginia at Fort Monroe with the intent of seizing Richmond. McClellan's advance stalled for a time when he encountered a much smaller Rebel force under John B. Magruder at Yorktown. For more, see Sears, *To the Gates of Richmond*, and Gallagher, *Richmond Campaign of 1862*.

44. In capable hands, the Springfield with its rifled barrel could strike a man hundreds of yards distant with lethal force, whereas the old smoothbore muskets were not of much account at ranges beyond a hundred yards. "The boys," James observed, "began to long for more stirring scenes than guarding railroad bridges." Bennett called the Springfield "the best gun I ever saw; and if we ever do get a crack at the rebels, they will get their bellies full, you can bet on it." James W. King, "Reunion of Michigan Veterans," unidentified newspaper, Aug. 25, 1873; "From the 11th Michigan Regiment," *Chronicle*, May 21, 1862.

45. Borrowing from "The Solitude of Alexander Selkirk" by William Cowper.

46. *Diary of Ira Gillaspie*, 25–26; Benjamin F. Wells to Susan M. Wells, May 2, 1862, Wells Papers; "Head Quarters Mich. 11th Reg't," *Reporter*, May 24, 1862; James Martin to his parents, May 6, 1862, Martin Letters.

3. The Celebrated John Morgan

1. Andrew Jackson, "Old Hickory," commanded the decisive victory over the British at New Orleans on the tail end of the War of 1812 and went on to serve two terms as the seventh U.S. president. James's stepmother Eliza (Van Buren) King's blood relationship to Jackson's vice president and longtime political colleague, Martin Van Buren, may have placed Jackson on a pedestal for the King family. After succeeding Jackson to the presidency, President Van Buren referred to Old Hickory as his "illustrious predecessor."

2. Three Rivers, Michigan, contains the confluence of the St. Joseph, Portage, and Rocky Rivers.

3. John Bell ran for president in 1860; he carried the states of Virginia, Kentucky, and Tennessee and finished fourth behind Lincoln, Douglas, and Breckinridge in the popular vote. Bell had served many years in Congress and was briefly secretary of war. Like many Southerners, he initially opposed secession but supported the Confederate cause once hostilities commenced. He fled Nashville upon the Federals' approach. Bell took up residence at his family's Stewart County, Tennessee, iron foundry after the war. Hoobler, *Guide to Historic Nashville*, 12.

4. Fort Zollicoffer, located on the Cumberland River west of Nashville, was abandoned upon the approach of the Federals. *Official Records of the Union and Confederate Navies*, ser. 1, vol. 22, pp. 675–76.

5. Andrew Ewing of Nashville served in the U.S. House of Representatives before the war. He was not in the Confederate Congress but was a judge on Braxton Bragg's military court in the Army of Tennessee. Upon his wartime death in Atlanta, Tennessee lost a gifted speaker and one of its most talented lawyers. Speer, *Prominent Tennesseans*, 139–40; Armstrong, *History of Hamilton County*, 45–46.

6. The 51st Ohio was stationed in Nashville on provost duty at this time.

7. Colonel Lewis was attending college in Lexington, Virginia, when the Revolutionary War broke out. He was blessed with a prolific extended family: during the war he raised and captained a company that counted twenty-two of his relatives in its ranks, and he would father eighteen children himself. Lewis was wounded twice at the Battle of King's Mountain. He survived the war and relocated to Nashville in 1793, attended the state's constitution convention in 1796, and was elected senator in the state legislature that year. Lewis, *Genealogy*, 99–101.

8. "Letter from Capt. Spencer," *Chronicle*, May 28, 1862; *OR*, vol. 10, pt. 2, p. 161; Benjamin F. Wells to Susan M. Wells, May 11 and 23, 1862, Wells Papers; "Columbia, Tenn., May 18, '62," *Reporter*, June 7, 1862; *Diary of Ira Gillaspie*, 27–28; Duke, *History of Morgan's Cavalry*, 163–64; "From the Eleventh," *Reporter*, June 21, 1862; Durham, *Nashville*, 159.

9. Benjamin Wells had written to his wife about James on May 11. "I have just given him some medicine. he has been sick for about 3 weeks but was getting much better until yesterday. today he has been quite sich he has Billious fever." Benjamin F. Wells to Susan M. Wells, May 11, 1862, Wells Papers.

10. The unit was not redeploying to Huntsville; they were merely ordered to escort a supply train in that direction. "Communicated," *Reporter*, June 7, 1862.

11. Lincoln's future vice president, Andrew Johnson, was a former U.S. congressman, senator, and governor of Tennessee, and one of the few prominent Southerners to remain loyal to the Union. Lincoln had appointed Johnson military governor of Tennessee in March. His companion, Neill Smith Brown, became in 1847 the youngest man ever elected governor of Tennessee. He served as U.S. minister to Russia from 1850 to 1853. A powerful public speaker, Brown possessed what one historian described as "matchless power of addressing crowds of men." The ex-governor said of his humble origins, "I had a native ambition to rise from obscurity and make myself useful in the world; to shine and be distinguished. . . . My poverty pushed me on." James mentions the 69th Ohio; it was a new regiment recently deployed from Camp Chase, Ohio. Speer, *Sketches of Prominent Tennesseans*, 8–9.

12. Brown's brother, John Calvin Brown, was captured leading a brigade at Fort Donelson. He would be released in a prisoner exchange in August, in time to take part in Bragg's upcoming Kentucky Campaign. After the war, John would follow in Neill's footsteps as governor of Tennessee. Neill Brown's sons, James Trimble and Neill S., both survived the conflict, the latter enduring four years of Confederate service. On this occasion, Brown spoke extemporaneously at the urging of friends, in an address lasting half

an hour. James captured the essence of the speech in his letter. The ex-governor claimed that he opposed secession from the beginning but stayed in his state "to moderate the fury of others." "I want this war stopped!" he exclaimed. "Old men die and their places are soon filled with others; but what a fearful chasm does the death of an army of noble young men make in society!" He warned that a continuation of the conflict must lead to the destruction of slavery. Phillips, *Governors of Tennessee*, 90–91; Speer, *Sketches of Prominent Tennesseans*, 9; "Unionism in Tennessee," *New York Times*, June 9, 1862.

13. A journalist accompanying Johnson described a similar speech the governor delivered at Murfreesboro a few days beforehand:

> The audience was a queer mixture of blue coats and butternuts. The latter stood listlessly inside the railing of the court-house yard. . . . But as Governor Johnson proceeded they began to exhibit more interest and attention. He seemed to know where and how to touch the hearts of the Tennesseans, and make them vibrate with patriotic emotions. In emphatic words he urged the deluded and erring Union men, who had by force or choice joined the rebel armies, to return to their allegiance. . . . It was a sight to observe the sway he seemed to have over them. . . . As a striking fact or forcible and pertinent illustration would present itself, they would burst into a laugh and applaud with approving cries.

Savage, *Life of Andrew Johnson*, 263, 266.

14. This heightened activity may have been triggered by rumors that John Morgan was plotting to capture Andrew Johnson. The artillery unit referred to is probably Battery B, 1st Ohio Light Artillery, under Captain William E. Standart. Savage, *Life of Andrew Johnson*, 264–65; *Diary of Ira Gillaspie*, 29.

15. Commissary Sergeant Elvah Pierce.

16. Quoting from "To Edward Noel Long, Esq." by Lord Byron.

17. Brigadier General Ebenezer Dumont was preparing a response to increasing Rebel activity in the area. Guerrilla and cavalry raids may have hindered the mail delivery. *Diary of Ira Gillaspie*, 29; Benjamin F. Wells to Susan M. Wells, June 20, 1862, Wells Papers.

18. James was transferred to the clerk position he described in his previous letter.

19. Dumont's expeditionary force consisted of several regiments of infantry, including the 69th Ohio, augmented with five hundred cavalry and ten guns. It was quite a novel experience for the Michiganders to march in such a large formation. Benjamin F. Wells to Susan M. Wells, June 10 and 20, 1862, Wells Papers; Mann, "Road to Murfreesboro," 82–85; *Diary of Ira Gillaspie*, 29–30.

20. Loosely quoting "The Closing Year" by George D. Prentice.

21. Verbal sparring between Union soldiers and Southern women was commonplace. Wiley, *Life of Billy Yank*, 107–8.

22. The Hermitage, Jackson's plantation and mansion, survives to this day as a historical museum. Chadwick noted that most visitors were being turned away in this period. "The Home of Andrew Jackson," *Reporter*, June 7, 1862.

23. Andrew Jackson authored the epitaph. Rachel died suddenly at age sixty-one, just weeks after her husband's election. James miscopied her age.

24. This armchair was described in 1880 as "cushioned with leather, which the relic-hunter has somewhat mutilated by snipping off pieces here and there." *Frank Leslie's Illustrated Newspaper*, Aug. 21, 1880, p. 419.

25. Major Andrew Jackson Donelson lived at the Hermitage for some time but did not own it. He was Rachel Jackson's nephew, raised by Andrew Jackson. He served as Jackson's aide-de-camp in the First Seminole War and was the president's private secretary throughout his White House years. Upon the birth of Donelson's daughter Mary in the White House, Jackson and Martin Van Buren were named as her godfathers. In 1856 the major ran unsuccessfully for vice president under Millard Fillmore. His younger brother, Daniel Smith Donelson, was the Confederate general who selected the site of Fort Donelson. Heiskell, *Andrew Jackson*, 3:273–74.

26. This is probably a vessel that was described as having "hoops, lid and handle . . . of silver, and the bottom was a magnifying glass, through which could be seen the joints of the vessel, which to the naked eye were invisible." Its inscription read, "This pitcher, consisting of 750 staves, made by James Cassidy, from a part of the elm-tree under which the treaty between William Penn and the Aborigines was concluded, that formed the state of Pennsylvania, was presented by the coopers of Philadelphia, to Andrew Jackson, President of the United States of America, December 1, 1834—as a testimony of their high estimation of his public services." Parton, *Life of Andrew Jackson*, 635; *Court Journal*, 60.

27. Duke, *History of Morgan's Cavalry*, 181–84, 195, 199, 202–3; James Martin to his parents, July 16 and 28, 1862, Martin Letters; Mann, "Road to Murfreesboro," 86–90; "The Eleventh Michigan after Morgan," *Chronicle*, July 23, 1862; *Diary of Ira Gillaspie*, 31–32; "From the Eleventh Michigan Regiment," *Chronicle*, July 30, 1862; *OR*, vol. 16, pt. 1, pp. 759–60, 762–63; Benjamin F. Wells to Susan M. Wells, Aug. 1, 1862, Wells Papers. (Wells in this letter is clearly confusing General Smith with Colonel Maxwell.)

28. John Hunt Morgan served as a lieutenant in the volunteer army of the Mexican War. In 1857 he organized a Kentucky militia unit. He was commissioned a Confederate captain in September 1861 and became colonel of the 2nd Kentucky Cavalry just in time for the Battle of Shiloh. Morgan rapidly earned a reputation for daring, highly successful raids behind Federal lines. For more, see Ramage, *Rebel Raider*.

29. James is probably referring to the poem "Hope and Memory," penned by George Crabbe and published in Baillie, *A Collection of Poems*.

30. The cavalry in question was led by Nathan Bedford Forrest. He captured Murfreesboro with its garrison and supply stores on July 13. Inaccurate news reports initially declared that the 11th Michigan was captured as well. "Important from the Southwest; Capture of Murfreesboro, Tenn., by the Rebels," *New York Times*, July 14, 1862; "The Eleventh Michigan after Morgan," *Chronicle*, July 23, 1862.

31. The notorious guerrilla Champ Ferguson was actually serving as a guide with Morgan at this time. Duke, *History of Morgan's Cavalry*, 182–83.

32. The quartermaster sergeant was John H. Underwood. President Lincoln had issued a call for 300,000 additional volunteers on July 2. This would be followed on August 4 with a requirement for the states to raise 300,000 militia on top of that. Regarding the foreign powers, the North feared British and French recognition of the Confederate government, which would clear the way for open European aid to the Southern cause. *Record of Service of Michigan Volunteers*, 11:96.

33. On July 11 Buell had ordered a detachment of the 11th to garrison Bowling Green; evidently James went there for some period of time between July 16 and July 29, and during that span he sent at least one letter to Jenny that has been lost. *OR*, vol. 16, pt. 1, p. 732.

34. The failure to cut off Morgan's retreat seems to have been the responsibility of Colonel Maxwell, who failed to advance at a key moment, rather than of General Smith. Benjamin Wells also blamed Smith but confused the general's role in this action with Maxwell's. "From the Eleventh Michigan Regiment," *Chronicle*, July 30, 1862; *OR*, vol. 16, pt. 1, pp. 759–60, 762–63; Duke, *History of Morgan's Cavalry*, 203; "From the Eleventh Michigan Regiment," *Chronicle*, July 30, 1862; Benjamin F. Wells to Susan M. Wells, Aug. 1, 1862, Wells Papers; James Martin to his parents, July 28, 1862, Martin Letters.

35. Duke, *History of Morgan's Cavalry*, 210–15; James Martin to his parents, Aug. 18, 1862, Martin Letters; "From the Michigan Eleventh," *Chronicle*, Sept. 3, 1862; *Diary of Ira Gillaspie*, 33–34; Robertson, *Michigan in the War*, 314.

36. Discipline in volunteer regiments was lax to begin with, and incidents of pillaging grew ever more frequent as the escalating human cost of the war hardened attitudes toward the enemy. One of the more notorious regiments in this respect was the 19th Illinois, destined to become brigade mates with the 11th Michigan. In response to guerrilla attacks, the Illinoisans had sacked the town of Athens, Alabama, in May. This growing spirit of vindictiveness was just as evident on the home front, where Northern newspapers railed against the resulting court-martial of the Illinoisans' colonel, John Basil Turchin. Ultimately, Turchin was not only reinstated but promoted as well. Haynie, *Nineteenth Illinois*, 166–68; Catton, *This Hallowed Ground*, 146–49.

37. Morgan's subordinate, Basil Duke (who was not present), later stated that the Federal cannon opened with canister, yet the Rebels suffered no casualties. He also had the assault occurring one day later and involving only sixteen Rebels. Surviving accounts of the same skirmish, penned by members of the 11th Michigan, asserted the accuracy of the Union cannon and musketry and consistently estimated the Rebel casualties between twenty-five and forty. Duke, *History of Morgan's Cavalry*, 215; James Martin to his parents, Aug. 18, 1862, Martin Letters; "From the Michigan Eleventh," *Chronicle*, Sept. 3, 1862; "From the Eleventh Regiment," *Reporter*, Aug. 30, 1862; *Diary of Ira Gillaspie*, 34; Robertson, *Michigan in the War*, 314; Monthly Return for August 1862, Regimental Service Records.

38. Benjamin Wells had feared for the life of his brother-in-law, Sergeant William Hoisington (also known as Wallace), and obtained permission to care for his sick comrade while the regiment was chasing its nemesis Morgan. A St. Joseph County history said of Hoisington, "He possessed the confidence and esteem of all his officers and men, and his death ... created a loss felt by all. He left a wife and one child to mourn the loss of a kind, indulgent husband and father." The sergeant's body could not be returned to Michigan due to the damage inflicted on the railroad tracks. Wallace's father, Drum Major Abishai Hoisington, was discharged from the service weeks later, officially due to rheumatism exacerbated by the recent hard marches. Benjamin F. Wells to Susan M. Wells, July 15 and Aug. 18, 1862, Wells Papers; *History of St. Joseph County*, 213.

39. The 11th was involved in the secret, and ultimately abortive, organization of a light brigade at this time. Its elements included the 11th Michigan and 19th Illinois, plus cavalry and artillery support. Overall command was assigned to Colonel John F. Miller,

with Stoughton in charge of the infantry. The foot soldiers would be wagon borne to maximize mobility, for the unit's primary purpose was to hunt down Morgan's and Forrest's cavalry. With priorities shifting due to Bragg's invasion, this force was disbanded promptly after marching to Murfreesboro and back. *OR*, vol. 16, pt. 2, pp. 430–32.

40. *Proceedings of Eighth Reunion*, 6; Benjamin F. Wells to Susan M. Wells, Aug. 29, 1862, Wells Papers.

41. James was not alone in forming this opinion. "For some time past," one of his comrades wrote anonymously to the *Three Rivers Reporter*, "there has been something of a 'skedaddle' by Buell's army from Alabama to this place, and it is looked upon by officers and soldiers as a most disgraceful act. It is said by those who ought to know, that he had men enough to have marched anywhere in Alabama or in Tennessee, and nothing but a gross and criminal inactivity during the Summer, has given place to the guerrillas in Tennessee, Alabama and Kentucky." "A Voice from Camp," *Reporter*, Oct. 25, 1862.

42. Nashville's makeshift fortifications consisted in part of walls of cotton bales studded with cannon. Andrew Jackson's men had experimented with such fortifications during the New Orleans Campaign of 1814–1815, but they discarded the idea when British artillery proved the 100 percent cotton defenses not only penetrable but disturbingly flammable as well.

43. For more about guerrilla warfare in Tennessee, see Ash, *Middle Tennessee Society Transformed*, 147–51.

44. James's older brother John, a clerk and salesman residing in Waukegan, Illinois, enlisted at the age of thirty as a private in Company G of the newly organized 96th Illinois Infantry on September 6. Partridge, *Ninety-Sixth Illinois*, 835.

45. President Lincoln's call on April 15, 1861 for 75,000 volunteers in the wake of the bombardment of Fort Sumter was welcomed with enthusiasm across the North. Some, however, criticized the president for not calling up more men. Stephen Douglas, for example, suggested the figure of 200,000 as more appropriate, but Lincoln was mindful of logistical realities, and of the possibility that summoning such a vast army could tip the political scales in the border states in favor of secession. Even so, many Confederates pointed to Lincoln's initial call for troops as the final straw in their decision to fight. Donald, *Lincoln*, 296.

46. Assistant Surgeon Arvin F. Whelan had resigned on October 13 and returned to Michigan. He would enlist as surgeon for the 1st Michigan Sharpshooters in January 1863. *Record of Service of Michigan Volunteers*, 44:96.

47. In early October, Stoughton led a foraging party of three regiments, including the 11th Michigan, into the hostile countryside. The column was ambushed by guerrillas three times during the march but brushed the enemy off with ease on each occasion. "I have the pleasure also of informing you," one Michigander wrote anonymously to the *Three Rivers Reporter*, "that we are not yet gobbled up by Morgan or 'Seceshia,' although for the present we are pretty closely hemmed in and our foraging parties have to be strongly guarded to protect them from marauders." Regarding the guerrillas, Basil Duke observed after the war that "the mountains of Kentucky and Tennessee were filled with such men, who murdered every prisoner that they took, and they took part, as their politics inclined them, with either side." Robertson, *Michigan in the War*, 314; "A Voice from Camp," *Reporter*, Oct. 25, 1862; Duke, *History of Morgan's Cavalry*, 183.

48. Rosecrans graduated fifth in the West Point class of 1842 and later taught there. At the outbreak of the war, he served briefly as aide-de-camp to George McClellan before receiving a brigadier general's commission. After early successes in western Virginia, including the Battle of Rich Mountain, he achieved promotion to major general in March 1862 and gained further command experience at the Battle of Iuka in September. Rosecrans repulsed an assault on Corinth, Mississippi, on October 4 and was selected as Buell's successor before the month was over. For more, see Lamers, *Edge of Glory*.

49. Company A's Private William H. Davis was a resident of Lockport Township, adjacent to Fabius. Subsequent letters show that Billy was on friendly terms with Jenny's family; in all probability he is the same individual referred to earlier by the nickname Odd. *Record of Service of Michigan Volunteers*, 11:27; 1860 United States Census, Schedule 1, Lockport, St. Joseph County, Mich., dwelling 1478, family 1494, Aug. 6, 1860; James W. King to Sarah J. Babcock, Apr. 23, 1863, King Collection.

50. Garrett could not be identified.

51. The Michiganders had not been paid in five months. James Martin to his parents, Nov. 29, 1862, Martin Letters.

52. Joshua Woodrow Sill graduated third in the West Point class of 1853 and later became an instructor there. He accepted the colonelcy of the 33rd Ohio in 1861, followed by promotion to brigadier in July 1862. Sill's West Point classmate Philip Henry Sheridan served on the frontier. He started out the war in quartermaster and commissary roles before securing appointment as colonel of the 2nd Michigan Cavalry in May 1862. He quickly earned promotion to brigadier and commanded a division with distinction at Perryville, foreshadowing the brilliant military career ahead of him. For more on Sheridan, see Wheelan, *Terrible Swift Sword*.

53. Norman, brother of the recently deceased William Hoisington, enlisted in the 13th Michigan Infantry on November 15, 1861, and was promoted to first lieutenant for bravery at Shiloh. Presumably the conversation dwelled on Norman's departed sibling, as James had witnessed the final stages of his bout with typhoid. *History of St. Joseph County*, 213.

54. McClellan was relieved of command on November 5. His timid advance on Richmond had been thwarted by Robert E. Lee in the Seven Days' Battles, enabling Lee to invade the North. McClellan stopped the Rebels on September 17 at the Battle of Antietam, the deadliest single-day clash of the war, but after he failed to pursue and engage Lee's retreating army, Lincoln's patience was exhausted. For more on McClellan, see Sears, *George B. McClellan*. For more on Antietam, consult Sears, *Landscape Turned Red*.

55. Rosecrans's army was divided into three wings. James Scott Negley commanded the 2nd Division of George H. Thomas's Center Wing. This division consisted of three brigades encompassing thirteen infantry regiments and three artillery batteries. Five of the regiments were Tennessean, four hailed from Ohio, and one each from Illinois, Indiana, and Pennsylvania. The 11th alone represented Michigan. James's count excludes the Tennessean brigade of James G. Spears, which remained behind to garrison Nashville. *OR*, vol. 20, pt. 1, pp. 174–82.

56. Horatio Phillips Van Cleve graduated from West Point in 1831 and served on the frontier until resigning five years later. He was commissioned colonel of the 2nd Minnesota in July 1861, rose to brigadier in March 1862, and commanded a division at Perryville.

57. Rosecrans's army was officially designated the 14th Army Corps, and his department was named the Department of the Cumberland. The name Army of the Cumberland was not officially adopted until January 9, 1863. Cozzens, *No Better Place to Die*, 14–15.

58. John Overton was the son of Judge John Overton, a founder of Memphis and close friend of Andrew Jackson. John Overton's son, John Jr., served on the staffs of Bushrod Johnson and Nathan Bedford Forrest. The Overton family land was confiscated by Union authorities. Speer, *Sketches of Prominent Tennesseans*, 176–77.

59. In an effort to cross the Rappahannock River at Fredericksburg, Virginia, unopposed, McClellan's successor Ambrose Burnside rushed the Army of the Potomac, but when the lead Union corps arrived, the pontoons required to traverse the river were nowhere to be found. Robert E. Lee's veterans were gifted the time needed to arrive and entrench across the river on Marye's Heights, a defensive paradise. Burnside fatefully decided to cross the river anyway and on December 13 suffered one of the war's most lopsided repulses. For more, see Rable, *Fredericksburg! Fredericksburg!*

4. The Worst Scourge

1. *Diary of Ira Gillaspie*, 42–43; James W. King to Sarah J. Babcock, Jan. 3, 1863, King Collection.

2. For more about the Battle of Stones River, see Daniel, *Battle of Stones River*. The best account of the 11th Michigan's tactical role in the battle is in Cozzens, *No Better Place to Die*. James W. King, "11th Michigan Infantry," *State Republican*, Jan. 11, 1901; *OR*, vol. 20, pt. 1, pp. 380, 420–21, 426–29, 688–89, 724, 763–65; *Diary of Ira Gillaspie*, 43–44; Haynie, *Nineteenth Illinois*, 186–89; Mann, "Road to Murfreesboro," 118–24; "From the Eleventh Regiment," *Reporter*, Jan. 17, 1863; Jim Lewis, "Battle of Stones River," 16, 25–26, 54.

3. "From the Eleventh Regiment," *Reporter*, Jan. 17, 1863; James Martin to his parents, Jan. 12, 1863, Martin Letters; Linus Squire to John Robertson, Dec. 22, 1863, in Regimental Service Records; *OR*, vol. 20, pt. 1, pp. 421–27, 429–30; Stevenson, *Battle of Stone's River*, 138–39; *Diary of Ira Gillaspie*, 44–45; Lewis, "Battle of Stones River," 48; Cozzens, *No Better Place to Die*, 203.

4. On the surface, this vengeful statement seems very much out of character for James, but it was common for a soldier's thoughts to turn to revenge after losing friends or family in battle. For more about combat's impact on Civil War soldiers, see Linderman, *Embattled Courage*, and Hess, *The Union Soldier in Battle*. "From the Eleventh Regiment," *Reporter*, Mar. 28, 1863; McPherson, *For Cause and Comrades*, 153–54.

5. Siblings Ezra, Anson, and Julia were the children of George Spencer and Eliza King's sister Atlanta. Both of James's parents would submit affidavits with the Spencer family's pension claim for Ezra's death. Anson, a Company A private, would

be discharged on May 28, 1863. Ezra Spencer's Pension Application, Application No. 13840, Certificate No. 13364, National Archives; 1850 United States Census, Schedule 1, Township of Lockport, St. Joseph County, Mich., dwelling 83, family 83, Aug. 31, 1850; *Record of Service of Michigan Volunteers*, 11:88.

6. Second Lieutenant Borden Mills Hicks of Company E recorded Joseph Wilson's fate. "Soon a stretcher was seen coming from the skirmish line, and our first man killed was brought in, a Lieutenant from Company 'F.' As we looked on his still form, we realized what war meant, our cheeks paled as we viewed our first sacrifice for our country." Hicks, "Personal Recollections," 525.

7. The 11th's casualties in the first day of fighting amounted to twenty-three dead, forty-eight wounded, and twenty-six missing. During the regiment's initial stand, which pitted them against Patton Anderson's and Alexander P. Stewart's brigades, Stoughton rode horseback, before the colors and under intense fire, until his steed was shot out from under him. Adjutant Chadwick's horse was hit repeatedly, and his sword was knocked from his hand by a solid shot. He climbed upon a borrowed mount, only to have that one shot up as well. Then his belt was sliced off by yet another bullet. Major Sylvester Smith was shot in the face, and, amid the smoke and chaos, was mistakenly carried into the Rebel lines by First Lieutenant Mathias Faulknor. Captain Oakes was not seriously harmed. The slain lieutenants included Joseph Wilson and Thomas Flynn. The latter was struck down during Negley's retreat. Mann, "Road to Murfreesboro," 119–20, 130; *Diary of Ira Gillaspie*, 44; "From the Eleventh Regiment," *Reporter*, Jan. 17, 1863.

8. The 7:00 A.M. January 1 shooting was mere skirmishing; New Year's Day did not see a general engagement. The afternoon assault on the train was no isolated incident. Bragg's superior cavalry arm enjoyed great success against the Federal supply line throughout the Stones River Campaign.

9. Charles W. Sanders's series of literature collections for students was published throughout the nineteenth century. The passage James goes on to cite appeared in *The School Reader: Fourth Book*. James quotes from a much longer passage that had previously been published in Dr. John Aikin's London magazine, *The Athenaeum*, in 1807, and attributed to A.L.B.: Aikin's sister Anna Laetitia Barbauld, the famous abolitionist, author, and controversial, outspoken peace advocate.

10. Loosely quoting Lord Byron's "Childe Harold's Pilgrimage."

11. Actually on New Year's Eve.

12. Quoting "Rain on the Roof" by Coates Kinney.

13. Babcock was not a lucky fellow: only sixteen men were drafted from all of St. Joseph County. An unwilling draftee could escape military service by convincing a doctor he was unfit to serve, by paying a fee known as "commutation," or by hiring another individual (a substitute) to serve in his stead. The name of Babcock's substitute is alternately recorded as Louis Gerrichtin, Girichton, or Lewis Gerieston. A recent immigrant from Bavaria, only fifteen years of age, Gerichten passed himself off as eighteen and joined Company I of the 1st U.S. Sharpshooters in April. Weeks later, he was shot in the right knee at Gettysburg. He was transferred to the Invalid Corps and successfully filed for a federal disability pension in April 1864. Gerichten later settled down in San Francisco, married a California native, and had at least three daughters. He is listed as a thirty-two-year-old gymnastics teacher in the 1880 census. He died

November 2, 1897. Captain Charles Stevens, in his otherwise meticulous history of Gerichten's regiment, mistakenly listed the substitute as killed in action. Robertson, *Michigan in the War*, 42; *Record of Service of Michigan Volunteers*, 44:140; Darius A. Babcock's Compiled Military Service File, National Archives; Louis Gerichten's Compiled Military Service File, National Archives; Louis Gerichten's Pension Index Entry, Application No. 43850, Certificate No. 32651, National Archives, accessed via fold3.com; 1880 United States Census, Schedule 1, San Francisco, Calif., p. 8, supervisor's district 1, enumeration district 83, dwelling 31, family 36, June 3, 1880; Stevens, *Berdan's United States Sharpshooters*, 344.

14. The promotion was effective March 1. This position demanded responsibility, integrity, and meticulous record keeping, for the regimental equipment and other property was necessarily scattered, and unscrupulous individuals might appropriate items for personal gain. James W. King's Compiled Military Service File, National Archives; Kautz, *Customs of Service*, 165–69.

15. James mentions Vicksburg and Grant throughout this chapter. For more on the Vicksburg Campaign, see Ballard, *Vicksburg*. Ulysses S. Grant graduated twenty-first out of thirty-nine cadets in the West Point class of 1843. Though twice brevetted in the Mexican War, he lost his way in the doldrums of peacetime service, resorted to alcohol, and resigned his commission. But he obtained a brigadier generalship early in the Civil War and gained nationwide fame by forcing the unconditional surrender of Fort Donelson. He was caught by surprise at Shiloh, however, and had not yet achieved the kind of success that would secure his reputation as the greatest of all Union generals. For more, see Smith, *Grant*.

16. The Peace Democrats, also known as Copperheads, were a vocal minority of Northerners, ranging from those advocating a more conciliatory approach to the Confederacy to those in outright opposition to the prosecution of the war. Their political clout ebbed and flowed with the fortunes of the Union war effort. The argument for peace had been fueled in recent months by military setbacks and opposition to the Emancipation Proclamation. The *Detroit Free Press* said of the proclamation, "No measure of this imbecile, malignant and tyrannical administration can compare to this," but most Union soldiers accepted the proclamation as a necessary war measure. Many soldiers' letters from this period are laced with hatred for the antiwar campaign. For an analysis of the peace movement's impact on Michigan, see Hershock, "Copperheads and Radicals," 29–69. (The *Free Press* quote above appears on p. 59.) For more on the Copperheads nationwide, see Weber, *Copperheads*. McPherson, *For Cause and Comrades*, 143–45; Mitchell, *Civil War Soldiers*, 127.

17. A portion of the Army of the Cumberland was stationed at Franklin under Major General Gordon Granger. On March 4 Colonel John Coburn led a reconnaissance of nearly two thousand bluecoats in the direction of Springfield. They collided with a much larger Rebel force under Earl Van Dorn at Thompson's Station and were surrounded. Most of the Federals were captured. *OR*, vol. 23, pt. 1, p. 7.

18. John King married Angeline Marie Simmons in Illinois on June 2, 1858. John King's Pension Application, Application No. 961025, Certificate No. 696827, National Archives.

19. Thomas Buchanan Read, renowned poet and painter, served on Lew Wallace's staff for part of the war.

20. Donelson and Shiloh have already been mentioned. Union forces managed to hang on to Pensacola, Florida, providing the navy with strategically located port facilities. For more, see Pearce, *Pensacola during the Civil War*. The Battle of Hartsville was actually a successful surprise attack carried out by John Hunt Morgan on December 7, 1862.

21. Referring to James Abram Garfield.

22. "The Oath" was read before Congress and was admired by President Lincoln. James submitted a very similar account of this Union meeting for publication in the *Three Rivers Reporter*. "If the Northern Copperheads," he wrote to the paper, "could have heard the words expressed by these men who had passed through blood and showers of iron for the sake of upholding the honor of our noble flag, they would have hid themselves in shame from the eyes of all honest men." Dickason, *Daring Young Men*, 27; "From the Eleventh Regiment," *Reporter*, Mar. 28, 1863.

23. The aunt from New York is probably Benjamin King's sister, Cornelia. Also mentioned is James's eldest sibling Henry, who was born in New York in 1830. Henry is listed in the 1863 draft registration with the "sight of one eye injured." He married Lucy Eggleston before the war and had two children by this time, Florence and Frank. They had three more offspring before Henry's death in 1872: James Arthur, Horace, and Cora. 1860 United States Census, Schedule 1, Fabius, St. Joseph County, Mich., dwelling 1009, family 1028, July 18, 1860; 1870 United States Census, Schedule 1, Fabius, St. Joseph County, Mich., dwelling 86, family 87, June 22, 1870; 1880 United States Census, Schedule 1, Fabius, St. Joseph County, Mich., dwelling 154, family 157, June 10, 1880.

24. Despite the draft's unpopularity back home, some soldiers were pleased that able-bodied men less patriotic than themselves would be compelled to join the war effort. "Bully for the President," James Martin declared. "It will make some of those fellows open their eyes that loves their freedom so well, but are not willing to fight for it. I am glad that I am not one of the men to be drafted. I would hate to have it said of me that I had to be drafted to fight for the old flag." Sergeant Benjamin Bordner of Company D said of the draftees, "Try their hand at thirteen dollars a month and put up as a mark to be shot at thats the fun of it … [otherwise] they will all get married and not let a woman for the soldier." James Martin to his parents, Mar. 6, 1863, Martin Letters; Benjamin F. Bordner to Mr. Austin and Sarah Hill, Mar. 3, 1863, Bordner Papers.

25. Lindley R. Harkness enlisted in Company E on August 9, 1862. He was promoted to commissary sergeant in March 1863. The sutler, Moore, could not be identified. *Record of Service of Michigan Volunteers*, 11:46.

26. Joseph Hooker graduated from West Point in 1837. He fought in the Second Seminole War and served on the frontier before accepting the position of adjutant at West Point. Hooker was thrice brevetted for gallantry in the Mexican War. He was appointed brigadier in May 1861, rose to division command in October, gained promotion to major general in May 1862, and then obtained corps command in time for South Mountain and Antietam. After carrying out Burnside's orders for the futile, repetitive charges at Fredericksburg, Hooker succeeded his superior as commanding general of the Army of the Potomac, in January 1863. For more, see Hebert, *Fighting Joe Hooker*. Lovell Harrison Rousseau, a native Kentuckian attorney and state senator, and state legislator of Indiana, was not a West Pointer but fought in the Mexican War. He was commissioned colonel and then brigadier in rapid succession in September

and October 1861. He led troops at Shiloh and Perryville, gained promotion to major general, and commanded a division at Stones River. For more, see Lee, *Kentuckian in Blue*. Kentucky lawyer Thomas Leonidas Crittenden served as an aide to Zachary Taylor in the Mexican War and was appointed consul to Liverpool, England, during Taylor's presidency. Commissioned brigadier general in late 1861, he commanded a division at Shiloh and earned promotion to major general in July 1862. He capably led Rosecrans's Left Wing at Stones River.

27. Escaped or freed slaves were referred to as "contrabands," in reference to the justification used by the Union for not returning the servants to their owners: as property, they could be confiscated as contraband of war. As the Union armies advanced, large numbers of slaves escaped servitude. At Murfreesboro this phenomenon was so significant that Rosecrans issued a general order delineating the manner in which they might be hired into army support roles: as teamsters, laborers, cooks, nurses, and officers' servants. Since March 13, 1862, Union officers had been prohibited from returning the slaves to their masters. But the same Union soldiers who served as the instrument of freedom generally looked upon the freedmen with contempt. The comments in this letter, deeply racist by modern standards, would not have turned heads in 1863. *OR*, vol. 23, pt. 2, pp. 17–18; McPherson, *Battle Cry of Freedom*, 497.

28. As mentioned earlier, Van Buren spent a year in the South before the war, seeking a change in climate due to health problems. He directed an academy near Yazoo City, northeast of Vicksburg, during that time.

29. The location of Ezra Spencer's burial place is a mystery. Presumably he lies either in his original burial location on the battlefield or was later reinterred at the Stones River National Cemetery, but without his body being identified. Nearly half the soldiers in the cemetery are listed as unknowns. For more about the issues surrounding the identification and burial of Civil War soldiers, see Faust, "Accounting: 'Our Obligations to the Dead,'" chap. 7 in *This Republic of Suffering*.

30. Another false rumor about Charleston. The birthplace of secession would hold out until February 1865. The city was targeted at this time by a joint army-navy operation.

31. Benjamin Grove Bennett, captain of Company D, had fought in the Mexican War and at Bull Run and was a former editor of the *Three Rivers Western Chronicle*. Pennsylvanian James Scott Negley fought in the Mexican War as well and was commissioned brigadier general in October 1861. He was promoted to major general on November 29, 1862. Timothy Robbins Stanley was elected to the Ohio senate before the war. He served as colonel of the 18th Ohio before achieving elevation to brigade command. *Record of Service of Michigan Volunteers*, 11:10.

32. Melvin Mudge of Quincy, Michigan, would be commissioned lieutenant colonel later in the month. *Record of Service of Michigan Volunteers*, 11:67.

33. James also submitted a description of the sword presentation to the *Three Rivers Reporter*, saying in part, "Few men are more deserving and entitled to greater praise than Colonel Stoughton. . . . Our regiment has never been praised as highly as some, but for all that Stoughton is a *bully* Colonel, and leads the *bulliest* regiment that ever went into a fight." "From the Eleventh Regiment," *Reporter*, May 2, 1863.

34. Hattie might be Jenny's first cousin Harriett, daughter of Peter and Artimetia (Blodgett) Blanchard. 1850 United States Census, Schedule 1, Constantine Township,

St. Joseph County, Mich., dwelling 702, family 705, July 29, 1850; 1860 United States Census, Schedule 1, Constantine Township, St. Joseph County, Mich., dwelling 907, family 923, July 14, 1860.

35. The 1860 census lists a thirty-six-year-old farmer by the name of Ansel W. Goodsel in Fabius. 1860 United States Census, Schedule 1, Fabius Township, St. Joseph County, Mich., dwelling 1098, family 1115, July 21, 1860.

36. James may have read the February 18 *Western Chronicle*, which covered a Michigan Democratic Party convention in Detroit and discussed Democratic victories in the St. Joseph County elections. The idea of comparing peace advocates with Tories was not a new one; *Lansing State Republican* newspaper editor D. C. Leach declared when the war broke out that "there will be but one party in the North, and that will be the Union party. If there is any other party it will be a Tory party." Thornton, *When Gallantry Was Commonplace*, 147–48; Martin J. Hershock, "Copperheads and Radicals," 45.

37. Rear Admiral Samuel du Pont had arrived off Charleston with a fleet of nine Federal ironclads. These vessels still retained a reputation for near invincibility, but that was quickly put to rest. Mines proved to be serious obstacles, but, more unexpectedly, the armored ships proved vulnerable to fort-based guns during a futile assault against Forts Moultrie and Sumter on April 7.

38. Van Dorn led a reconnaissance-in-force against Major General Gordon Granger at Franklin on April 10. The Rebels withdrew after a relatively minor clash.

39. Privates Kneeland Latham of Company E and Jonathan Ferguson of Company I were nabbed while in pursuit of some errant army mules. "Communicated," *Reporter*, June 7, 1862.

40. Alexander McDowell McCook graduated from the West Point class of 1852 and served on the frontier and as a West Point tactics instructor before the war. Commissioned colonel immediately following the attack on Fort Sumter, he fought at Bull Run and secured the rank of brigadier general in September 1861. After leading a division at Shiloh, he was promoted to major general and led a corps at Perryville. At Stones River he was caught by surprise and witnessed the rout and utter humiliation of two of his divisions before his third division commander, Phil Sheridan, blunted the Confederate assault. For more, see Fanebust, *Major General Alexander M. McCook, USA*.

41. Private William Bournes, an Irishman serving in Company C, was a friend of the Babcock family. *Record of Service of Michigan Volunteers*, 11:13.

42. James refers to Second Lieutenant James M. Whallon of Company C. No additional information could be found regarding the wedding. *Record of Service of Michigan Volunteers*, 11:101.

43. This should read the thirtieth of December. Stanley directed the 19th's skirmishers to clear the Rebels from several buildings and from a brick kiln owned by Giles Scales Harding, whose house James describes visiting later in this letter. *OR*, vol. 20, pt. 1, p. 425; Coleman, "Harding House," 1, 5.

44. Joshua Sill was killed leading his brigade the morning of December 31. He was slain wearing Phil Sheridan's coat, which he had mistaken for his own.

45. James and his comrades were visiting the two-story log house of Giles and Mary Harding, site of the brigade's skirmish on December 30. Presumably Mary was the controversial pianist, and she had cause to be ornery. Her house had been

commandeered for use as a Union hospital for a time, and her chickens, geese, and horses were stolen. Far worse than that, some soldiers had tried to hang her husband—officers intervened at the last moment to save him. One historian described her as "a small lady and very spunky." The three other ladies mentioned by James may have been Harding daughters Mary, Elizabeth, and Ellen Amy. After the war, some ex-bluecoats returned to see the house, photographed the piano, and had a local printer create postcards of it. James also mentions First Lieutenant Ephraim Gaylord Hall of Company F, an intellectual giant known as "the Walking Encyclopedia" of the 11th Michigan. Hall was shot through the neck during the retreat at Stones River. The bullet passed near the base of his brain yet he miraculously survived, only to suffer capture and confinement in Libby Prison. Hall's reappearance with the 11th indicates his release in a prisoner exchange. Coleman, "Harding House," 1–2, 4–5, 13; "Capt. E. G. Hall," 3.

46. *Frank Leslie's Illustrated Newspaper* published many illustrations of battle scenes.

47. In mid-April, Admiral David Porter's fleet successfully ran past the guns at Vicksburg, securing sufficient naval support south of the city to ferry Grant's troops across the Mississippi. The Federals crossed the river on April 30 and defeated a Rebel force at Port Gibson the next day. Grant then boldly struck inland. On May 14 his army smashed Confederate reinforcements gathering at Jackson, Mississippi, under Joseph E. Johnston. When Lieutenant General John Pemberton belatedly advanced from Vicksburg to support Johnston, Grant's army turned and soundly thrashed the butternuts at Champion Hill on the sixteenth, and again at Big Black River the next day. Pemberton retreated to the safety of Vicksburg's fortifications.

48. The heavens were not to blame for the Union disaster at Chancellorsville. Joseph Hooker accomplished unopposed crossings of the Rappahannock and Rapidan Rivers, placing the bulk of his army on Lee's left, but then he lost his nerve, and with it the initiative. Stonewall Jackson seized the resulting opportunity to hit the Army of the Potomac in the flank. Jackson was mortally wounded by friendly fire, but Hooker was forced to retreat back across the river with severe losses, setting the stage for Lee's second invasion of the North. For more, see Sears, *Chancellorsville*.

49. A false rumor. Grant assaulted Vicksburg's defenses on the nineteenth and twenty-second of May, but gained little more than a lengthening casualty list. Unable to break the Rebel defenses, the Federals settled in for a siege.

50. William A. Selkirk was convicted by military commission for the murder of Adam Weaver of Wilson County. It was Colonel Stoughton's duty as provost marshal to oversee Selkirk's execution. A *New York Times* reporter related the scene:

The prisoner seemed to suffer no agitation of any consequence, until he appeared in sight of the gallows, when his countenance suddenly put on a death-like pallor, and only regained the calm serenity it wore when the wagon passed through the immense multitude that awaited the arrival of the prisoner. The execution was witnessed by at least 10,000 people, mostly soldiers. Immediately in front of the prisoner stood the son and daughter of the murdered man; the latter actually made application to the authorities for permission to adjust the rope about the neck of the doomed man. The prisoner asserted his innocence most positively just previously to being swung off, and said he had hopes of eternal life.

James refers to copying Selkirk's confession, but that document, in James's handwriting, survives and is actually an assertion of innocence. Selkirk claimed the murder was committed by another man while Selkirk and a third individual by the name of Rhodes stood nearby. Rhodes, on the other hand, told the 11th Michigan's chaplain that he and Selkirk were six miles away at the time. "A Military Execution in Gen. Rosecrans' Army," *New York Times*, June 14, 1863; Union Provost Marshal's File of Papers Relating to Individual Civilians, 1861–67, microfilm publication M345, roll 241, National Archives.

51. Joseph could not be identified.

52. Many Union soldiers commented on the startling prevalence of tobacco use among Southern women. Wiley, *Life of Billy Yank*, 101–2.

53. The Rebels had tested the defenses at Franklin again on June 4. OR, vol. 23, pt. 1, pp. 359–62.

54. The Confederate forces in Mississippi included a number of infantry and artillery units from South Carolina. It was not unusual for eastern units to serve in the West, and vice versa.

55. Jenny's first cousin John Milo Hammond was a son of William and Laura (Blodgett) Hammond.

56. Ezra Blodgett Fulkerson, the brother of Jenny's deceased cousin Alden, moved to California. He would join the 2nd California Infantry in July 1864. He survived the war. Orton, *Records of California Men*, 466.

57. Grant maintained his stranglehold on Vicksburg. Food shortages ensued, and the city was relentlessly bombarded. Joseph Johnston was tasked with rescuing Pemberton's defenders, but his force was inadequate to the task, and Vicksburg remained isolated from the rest of the world. Meanwhile out east, subsequent to the spectacular Rebel victory at Chancellorsville, the Army of Northern Virginia launched its second invasion of the North. Hooker, unsure of Lee's intentions, was initially slow to pursue.

5. This Cannot Be a Defeat

1. For more on the Tullahoma Campaign, see Bradley, *Tullahoma*. OR, vol. 23, pt. 1, pp. 447–48; Bradley, *Tullahoma*, 86.

2. Belknap, *Michigan Organizations*, 109–10.

3. John T. Wilder's mounted Federal infantry, armed with Spencer repeating rifles, overpowered the Rebel defenders of Hoover's Gap with superior firepower. James refers to the battle fought at the pass of Thermopylae in 480 B.C., in which a vastly outnumbered Greek army held off a Persian assault for days, due in large part to superior defensive terrain.

4. Joseph Jones Reynolds graduated tenth in his West Point class of 1843 and later served the academy as an instructor. He was appointed brigadier general soon after the war broke out and achieved a victory at Cheat Mountain in September 1861. With promotion to major general in November 1862, Reynolds now commanded a division.

5. For more about Robert E. Lee's disastrous defeat at Gettysburg, Pennsylvania, see Guelzo, *Gettysburg*. On July 4, the very day of Lee's retreat, Vicksburg surrendered its garrison of thirty thousand troops to Ulysses S. Grant. Five days later, the Confederate Mississippi River stronghold of Port Hudson capitulated as well. With Federal navigation

of the Mississippi restored, a critical route of trade, communication, and supply was reopened, the Confederacy was split in two, and Grant's army was freed up for operations elsewhere. The claim about Charleston, however, was another false rumor.

6. The 96th Illinois had been on garrison duty at Wartrace since July 3. *History of Jo Daviess County*, 401.

7. John King was detailed for detached quartermaster duty. The 96th Illinois departed Wartrace for Elk River the same day this letter was written. *History of Jo Daviess County*, 401.

8. Addison T. Drake was the 11th's regimental quartermaster, but starting in February 1863 he bounced between roles as acting brigade, division, and corps quartermaster. These assignments probably prompted James's promotion to quartermaster sergeant. The former quartermaster sergeant, Lieutenant John H. Underwood, was acting quartermaster of the 11th in Drake's absence. James, as quartermaster sergeant, worked closely with both Drake and Underwood. Monthly Returns, Regimental Service Records; *Record of Service of Michigan Volunteers*, 11:30, 11:96.

9. The 11th crossed the river at Caperton's Ferry; James and the regimental train, at Bridgeport. Belknap, *Michigan Organizations*, 109.

10. The missing soldiers, First Sergeant Elmer Bradley and Corporal Oliver W. Brockway of Company K, and Private James H. Ensign of Company A, were captured and later perished at Andersonville. *OR*, vol. 30, pt. 1, pp. 271–73, 284–85, 292–302, 326–28, 367, 376–77, 384; Belknap, *Michigan Organizations*, 110–12, 243–45; Evans, *Battle of Davis' Crossroads*, 30–41; Haynie, *Nineteenth Illinois*, 215–16; Robertson, "Chickamauga Campaign, McLemore's Cove," 22, 45; James W. King, "11th Michigan Infantry," *State Republican*, Jan. 11, 1901; *Record of Service of Michigan Volunteers*, 11:13–14, 11:32–33, 11:50.

11. Stanley's brigade had been weakened by the detachment of the 69th Ohio. For more about the Battle of Chickamauga, see Cozzens, *This Terrible Sound*, and Powell, *Maps of Chickamauga*. *OR*, vol. 30, pt. 1, p. 329; Belknap, *Michigan Organizations*, 246.

12. A modest stone monument to the 11th Michigan, nestled in the woods west of Lafayette Road, marks the location of General Adams's capture. During Gracie's assault, the Michiganders were attacked by the 2nd Alabama Battalion and the 43rd Alabama Infantry. Salling, *Lousianians*, 125–31; Belknap, *Michigan Organizations*, 113–19, 245–50; James Martin to his parents, Sept. 24, 1863, Martin Letters; *OR*, vol. 30, pt. 1, pp. 329–30, 379–82; James W. King, "11th Michigan Infantry," *State Republican*, Jan. 11, 1901; Robertson, "Chickamauga Campaign, Battle," 21–23; Hicks, "Personal Recollections," 528–31; Powell, *Maps of Chickamauga*, 156–57, 164–65, 209, 226–27, 230–31; Haynie, *Nineteenth Illinois*, 220–38; *OR*, vol. 30, pt. 2, pp. 304–5, 415–18, 423–28, 504–5.

13. Daniel Rose to his mother, Sept. 27, 1863, Masterson Collection, 1861–1865, Western Michigan University Archives and Regional History Collections (hereafter cited as Masterson Collection); James W. King to Sarah J. Babcock, Oct. 4, 1863, King Collection; Hicks, "Personal Recollections," 530–31; *OR*, vol. 30, pt. 1, pp. 172, 382; James W. King, "11th Michigan Infantry," *State Republican*, Jan. 11, 1901.

14. Newberry was killed in a brief but sharp clash with the 7th South Carolina immediately after the 11th arrived on Horseshoe Ridge. Belknap, *Michigan Organizations*, 116–17; Powell, *Maps of Chickamauga*, 209.

15. In the immediate aftermath of Chickamauga, the War Department ordered Ulysses S. Grant to reinforce Chattanooga. Grant responded by sending Sherman from Vicksburg. Additionally, the 11th and 12th Corps of the Army of the Potomac were dispatched under Joseph Hooker to the same purpose.

16. Interestingly, James was not alone in declaring victory. "I presume," James Martin wrote his parents, "some of the northern doughheads will [say] that Rosecrans got out generaled but I dont think so, he started to take Chatanooga, and we are here and I think that we are able to stay." Unlike much of the Federal army, Stanley's brigade had dealt out far more punishment than it received at Chickamauga, and their confidence must have been elevated far above that of the Army of the Cumberland as a whole. Captain David Bremner of the Nineteenth Illinois agreed with his comrades in the 11th: "The Battle of Chickamauga successfully closed General Rosecran's campaign for the capture of Chattanooga. I say successfully, because it was the objective of the campaign, and the Confederates were defeated in every attempt to drive back Rosecrans and prevent the concentration of his army at Chattanooga." James Martin to his parents, Sept. 24, 1863, Martin Letters; Haynie, *Nineteenth Illinois*, 242.

17. George Henry Thomas graduated twelfth in the West Point class of 1840, won two brevets in the Mexican War, and taught at West Point before joining the 2nd U.S. Cavalry. He was a rare professional soldier of Virginia who sided with the Union. Commissioned brigadier general in August 1861, Thomas claimed one of the early Union victories, at Mill Springs; led a division at Shiloh and received promotion to major general; served as second in command under Buell at Perryville; and led Rosecrans's Center Wing at Stones River. His legendary stand on Horseshoe Ridge earned him the nickname "Rock of Chickamauga," but his military career never garnered the respect it deserved. For more, see Wills, *George Henry Thomas*.

18. The Battle of Davis's Crossroads is not mentioned in James's surviving letters, probably indicating missing correspondence predating Chickamauga. Given the chaotic state of affairs in that period, it is quite possible that the letter in question never reached Michigan.

19. James Longstreet arrived with reinforcements from Virginia just in time to bolster Bragg's army at Chickamauga.

20. As discussed in a previous note, the success of Stanley's brigade at Chickamauga seems to have imbued his men with a level of confidence that did not necessarily pervade the Army of the Cumberland as a whole. Although there was a general awareness that Bragg took heavy casualties, most of the Federals could not possibly have shrugged off the fact that the battle had ended in a Union rout followed immediately with siege conditions at Chattanooga.

21. The depot at Bridgeport was key to supplying the army in Chattanooga. Confederate Joseph Wheeler embarked on a cavalry raid behind the Union lines on October 1. The next day, he intercepted a Federal supply convoy of eight hundred wagons. The small guard was quickly overwhelmed.

22. Sergeant Major Washington Irving Snyder was wounded shortly after the unit arrived on Horseshoe Ridge. The 11th briefly traded volleys with the 7th South Carolina before the Rebels fell back. Belknap, *Michigan Organizations*, 116–17; Powell, *Maps of Chickamauga*, 209.

23. The Confederates retaliated later in the day, as narrated by James Martin. "They threw shell and shot over us, and under us, and right in amongst us, but dident happen to hurt any one. . . . A thirty two pound solid shot just missed our heads and struck about ten feet in front of us. I tell you, it makes a fellows hair stand up to have as large a ball as that just miss his old head." James Martin to his father, Oct. 15, 1863, Martin Letters.

24. Cozzens, *Shipwreck of Their Hopes*, 20.

25. James W. King, "The Charge of Mission Ridge," *Three Rivers (Mich.) Tribune* (hereafter cited as *Tribune*), May 1, 1896, photocopy in the King Collection.

26. Thomas had successfully proposed a picket ceasefire to Braxton Bragg. James Martin said of the picket scene, "Our pickets and the rebel pickets stand about 40 rods apart in plain sight of each other, but both parties have orders not to fire on the other. We exchanged papers with them and talked with them about the battle and had a regular old conflab." A few days later, Martin reported relations growing even warmer. "The rebels are as friendly on picket as ever, our pickets and the 'rebs' isent only about ten rods apart where we are now, both parties get water at the same spring; they exchange papers, trade knives and tobacco and blag guard [blackguard] one other like the devel." Cozzens, *Shipwreck of Their Hopes*, 49–50; James Martin to his father, Oct. 15 and 26, 1863, Martin Letters.

27. A false rumor.

28. James mentioned in his April 23 letter that Lemuel Pierce was serving as division train master. It was a daunting role to fill, under unrelenting pressure to keep the besieged army fed.

29. Regarding Hooker and Longstreet, James refers to the Battle of Wauhatchie, fought on the night of October 28–29, in which the Federals successfully defended their recently secured and desperately needed supply route, dubbed the Cracker Line.

30. Major General Ambrose Burnside had occupied Knoxville. Isaac Stell was wounded at Port Hudson on June 19, and was ultimately discharged as a result. His brother Benjamin was the future husband of Jenny's first cousin, Elizabeth Jane Blodgett. He would be wounded in the victory over James Longstreet at Fort Sanders on November 29. He later transferred to the 17th Michigan and served out his full term of enlistment. *Record of Service of Michigan Volunteers*, 2:165, 6:136.

31. Daniel Webster Holbrook was discharged on October 28. It was Holbrook who first obtained a clerk position for James, whose service in that capacity likely led to his promotion to quartermaster sergeant. *Record of Service of Michigan Volunteers*, 11:49.

32. William Tecumseh Sherman was approaching Chattanooga with additional reinforcements. Combined with Hooker's prior arrival, the number of Union soldiers at Chattanooga would fully double. Preparing to leverage his advantage, Grant ordered Thomas to conduct a reconnaissance in force toward Missionary Ridge on November 23. Thomas John Wood's division overran the Confederates on Orchard Knob, a sharp rise located between the city and the ridge. Grant informed Henry Halleck that the stage was set for a decisive battle.

33. Note that James's Medal of Honor nomination file is cited here; he would be recommended for the medal in 1901 but was ultimately rejected. Further discussion follows in chapter 8 and appendix A. James W. King's Medal of Honor Nomination File, Record and Pension Office, Document No. 670252, National Archives; James W. King to J. C. Burrows, Feb. 17, 1902, King Collection.

34. James charged with Hicks's Company E, which had been raised in Three Rivers. At least three secondary sources—Durant, Dudd, and *Three Rivers*—suggest or state that James was the first Federal to crest Missionary Ridge, but eyewitnesses consistently used phrases like "one of the first" or "among the first." James W. King's Medal of Honor Nomination File, Record and Pension Office, Document No. 670252, National Archives; James W. King to J. C. Burrows, Feb. 17, 1902, King Collection; Belknap, *Michigan Organizations*, 120–22; Robertson, *Michigan in the War*, 318–20; James W. King, "The Charge of Mission Ridge," *Tribune*, May 1, 1896; Hewett et al., *Supplement to Official Records*, vol. 6, pt. 1, pp. 106–9, 116–17, 125–26, 135–37, 141–50; *OR*, vol. 31, pt. 2, pp. 479–89; Hicks, "Personal Recollections," 532–34; "Letter from the Eleventh Mich.," *Reporter*, Dec. 12, 1863; Durant, *History of Ingham*, 160; Lillian Shafer Dudd, King Genealogy, Dudd Collection; *Three Rivers*, 27.

35. There seems to be no other surviving reference to the friendship between James and Colonel Wallace W. Barrett. For an overall account of the Battle of Missionary Ridge, see Cozzens, *Shipwreck of Their Hopes*. However, the fight between Stoughton's and Strahl's brigades is best understood by examining recently published primary sources in Hewett, *Supplement to Official Records*, as cited in the adjacent notes, in conjunction with the cited Federal accounts in *OR*. James's role in the battle is further fleshed out in chapter 8 and appendix A. Belknap, *Michigan Organizations*, 121.

36. The opposition at the ridge summit had consisted of the 19th and 24th Tennessee Infantry, with artillery support from the Eufaula Battery and Stanford's Battery. The troops that fled the crest upon James's approach were a rallied remnant of a Floridian brigade. The Confederate after-action reports of all five officers commanding infantry on either side of the gap—from the regimental to division level—unanimously declared that the breech James charged into triggered their troops' precipitous retreat. Regarding James's reaction to his injury, the potential loss of romantic relationships was the single greatest fear facing amputees. American society, however, considered it a woman's duty to stay true to a wounded soldier. Hewett, *Supplement to Official Records*, vol. 6, pt. 1, pp. 106–11, 116–17, 125–26, 137, 147–48; James W. King to J. C. Burrows, Feb. 17, 1902, King Collection; James W. King to Sarah J. Babcock, Feb. 21, 1865, King Collection; Nelson, *Ruin Nation*, 190–93.

37. The motivation behind Darius Babcock's enlistment is a mystery. Less than one year prior, he had paid a substitute in order to escape army service. What changed his mind? Had his health improved? Was he enticed by the volunteer bounty? It is interesting to note that he enlisted just as James completed his furlough. Perhaps the young man's return home as a hero of Missionary Ridge sparked a sense of patriotism, or a yearning for glory, in the forty-three-year-old enlistee. James W. King to Sarah J. Babcock, Mar. 6, Apr. 26, 1864, and Feb. 21, 1865, King Collection; James W. King, "Stranger Than Romance," 2, King Collection; James W. King, "Truth Is Stranger Than Fiction," 3–4, King Collection; Partridge, *Ninety-Sixth Illinois*, 835–36; James W. King's Compiled Military Service File, National Archives; Susan M. Wells to Benjamin F. Wells, Feb. 8, 1864, Wells Papers; Darius A. Babcock's Compiled Military Service File, National Archives; *Record of Service of Michigan Volunteers*, 6:8.

38. James frequently refers to Daniel Webster Holbrook as Webb from this point forward. Holbrook had been discharged from the 11th for disability in October 1863, but James's next letter places him in Chattanooga. Later letters suggest that Holbrook

had obtained a civilian job in a regular army assistant quartermaster's office by this time. James W. King to Sarah J. Babcock, Feb. 13, 1865, King Collection.

39. Addison Drake was the corps's acting assistant quartermaster at this time; John H. Underwood continued to fill his shoes as acting regimental quartermaster. Monthly Return for Mar. 1864, Regimental Service Records.

40. The regiment employed a number of ex-slaves in 1864. Some of them would accompany the 11th back to Michigan when the unit mustered out. Hicks, "Personal Recollections," 542.

41. The regiment had rejoined its brigade at Graysville, Georgia, on March 15. *Annual Report Adjutant General 1864*, 148; Daniel Rose to his mother, Mar. 20, 1864, Masterson Collection.

42. Quoting from John Finley's "Bachelor's Hall."

43. Two hundred forty-seven veterans of the 6th Michigan Heavy Artillery had traveled back to Michigan for their thirty-day reenlistment furlough. Robertson, *Michigan in the War*, 267.

44. Jenny's cousin Augustus Milo Wellman, son of James Henry Wellman and Alvira Blodgett, was among a batch of recruits who arrived from Michigan. Wellman had enlisted on January 18. *Record of Service of Michigan Volunteers*, 11:100; Muster and Descriptive Rolls, Regimental Service Records.

45. Wounds were seen as proof of courage, and the desire to go on fighting after suffering injury in battle further served as proof of patriotism. Elliott's diagnosis of James's wound was ankylosis of the elbow. James W. King, Army of the United States Certificate of Disability for Discharge, Apr. 11, 1864, photocopy in the King Collection; James W. King's Pension Application, Application No. 125139, Certificate No. 87706, National Archives; Nelson, *Ruin Nation*, 179–82.

6. The Cannons' Deep Roar

1. Daniel Rose leaned against reenlisting because all three of his brothers had enlisted as well, leaving his widowed mother shorthanded at home. When Benjamin Wells wrote jokingly to his wife of reenlisting, she replied that "I fear my wrath would rise a little and I might tell you to enlist for life while you were at it. I am afraid I would be tempted to hunt me up another man if you were to make up your mind to stay three years longer. . . . I should surely think you did not care as much for us as I thought you did." It is difficult, however, to pinpoint why the men of the regiment as a whole viewed reenlistment so much more negatively than most Union army units. *Annual Report Adjutant General 1864*, 6, 150, 549; Thornton, *When Gallantry Was Commonplace*, 208; Daniel Rose to his mother, Feb. 4, 1864, Masterson Collection; Susan M. Wells to Benjamin F. Wells, July 8, 1864, Wells Papers.

2. Sherman in fact was busy stockpiling supplies in Nashville, securing a sufficient flow of stores from there to Chattanooga by rail, and gathering and preparing his combined armies for the military campaign ahead. By the end of April Sherman would marshal close to 100,000 soldiers, abundantly supplied and ready to march.

3. The arming of blacks was controversial in the North, and Southerners were seething, deeming this act an uncivilized outrage. On April 12, the same day James began

penning this letter, Confederate troopers under Nathan Bedford Forrest massacred black Federal soldiers upon capturing Fort Pillow in Tennessee. For more about the U.S. Colored Troops and the white officers who commanded them, see Glatthaar, *Forged in Battle*.

4. Loosely quoting T. S. Arthur's "Do They Miss Me?"

5. Darius Babcock departed Michigan with his regiment on April 30, bound for Port Hudson, Louisiana, by railroad. Near Effingham, Illinois, an axle on the train snapped, and the cars jerked violently as the train derailed. Babcock suffered a serious back injury but continued on to Louisiana with his regiment. Darius A. Babcock's Pension Application, Application No. 334874, Certificate No. 373209, National Archives.

6. Lyrics from "My Own Native Land" by William Bradbury.

7. Lemuel Pierce would be discharged for disability on May 6. *Record of Service of Michigan Volunteers*, 11:73.

8. Preacher, orator, and abolitionist, Henry Ward Beecher was the brother of Harriet Beecher Stowe, author of *Uncle Tom's Cabin*.

9. The two halves of the shelter tent were fastened together to provide lodging for two men.

10. Each regiment was allowed one ambulance and one wagon, and personal belongings were kept to a bare minimum. However, the army as a whole brought fifty wagons per thousand men, making this proportionately the second largest supply convoy of the war. Wagons belonging to the brigade, division, corps, and army trains were restricted to the absolute necessities of clothing, ammunition, and food but carried these in great abundance. Sherman was balancing mobility with independence from his supply line. Castel, *Decision in the West*, 117.

11. Governor John Brough of Ohio was the driving force behind the raising of regiments under hundred-day enlistments. These troops would guard railroads and installations away from the fighting, freeing combat-ready regiments for front-line duty. Many of these short-time soldiers guarded the railroad behind Sherman's upcoming advance and thus helped to maintain troop strength for the offensive. The hundred-day men would further prove their worth when Jubal Early threatened Washington, D.C., in July. For more, see Leeke, *A Hundred Days to Richmond*.

12. In addition to the Army of the Cumberland, Sherman's command included James B. McPherson's Army of the Tennessee (not to be confused with their similarly named Confederate counterparts) and John M. Schofield's Army of the Ohio. The 19th Illinois split paths with the 11th Michigan just before the outset of the campaign, transferring to Absalom Baird's division. For more on the Atlanta Campaign, see Castel, *Decision in the West*. Johnson, *Body of Brave Men*, 486–87.

13. "We stood and waited nearly all day long," wrote Borden Hicks of the June 27 assault on Kennesaw Mountain, "for the command to charge, the works in our front were not over four rods apart, and we knew it meant death to the charging column. It was the most trying day that we experienced in our whole term of service." For more on the extended confrontation at Kennesaw, see Hess, *Kennesaw Mountain*. *OR*, vol. 38, pt. 1, pp. 560–61, 570; Castel, *Decision in the West*, 204–5; Hicks, "Personal Recollections," 539–40.

14. On May 14 at Resaca, the 25th Michigan assaulted the enemy works, crossing an open field and wading through a waist-deep creek under fire. The Rebels were driven

from their fortifications, but the 25th absorbed fifty casualties in less than five minutes. On the following day, the 19th Michigan charged a Rebel gun battery, assisting in its capture at a cost of fourteen killed and sixty-six wounded, including Colonel Henry C. Gilbert, mortally wounded. Robertson, *Michigan in the War*, 394, 458.

15. The recently promoted Second Lieutenant Cuthbert Dixon was wounded on May 27, likely in the same artillery barrage that injured division commander Richard Johnson. Dixon recovered to see out the balance of his enlistment. *Record of Service of Michigan Volunteers*, 11:29; William E. Raymond to Duke, June 5, 1864, William E. Raymond Papers, 1863–1864, Bentley Historical Library.

16. The quote is from the George Washington Cutter poem "E. Pluribus Unum."

17. Darius Babcock was admitted to St. Louis Hospital in New Orleans from June 8 to June 19 due to the back injury he sustained on the train ride south. Unable to perform strenuous duty, he would serve out the balance of his enlistment detailed as a cook. His regiment was posted at Morganza, Louisiana, for most of June. Darius A. Babcock's Pension Application, Application No. 334874, Certificate No. 373209, National Archives; Robertson, *Michigan in the War*, 267.

18. The Confederates had launched a failed assault at Kolb's Farm on the twenty-second, but the 11th was not involved.

19. John King and the 96th Illinois's brigade were relieved from the front line by Stoughton's men on the twenty-second. The 96th had been heavily engaged on several occasions during the campaign, particularly at Rocky Face Ridge, Resaca, and in the heavy skirmishing before Kennesaw Mountain. Stoughton's inherited position was a mere hundred yards from the Confederate works, placing his men under constant threat from sharpshooters and artillery. A Rebel gun battery lay straight ahead, less than 250 yards distant. *OR*, vol. 38, pt. 1, pp. 561, 571.

20. The soldiers had already suffered much in this campaign, dodging bullets and shells day after day, and now their morale took another blow: mail service became spotty in both directions for weeks. Just enough letters trickled through to lead both the bluecoats and their correspondents back home to the conclusion that mail was getting through just fine, and that they were simply being neglected in this most trying time. When Benjamin Wells complained to his wife about the lack of letters, Melissa replied, "If you had recd them you would not feel as you do towards me nor be so impatient with me. What can I do I certainly can do no more than write them and mail them." Susan M. Wells to Benjamin F. Wells, 22 July 1864, Wells Papers.

21. Unfortunately, this drawing could not be located.

22. Private William Mansion had enlisted in Detroit in August 1861. He was wounded on June 29 and died the next day. *Record of Service of Michigan Volunteers*, 11:62

23. The 6th Michigan had been transferred from Port Hudson to Vicksburg in late June to serve with an engineer brigade. Robertson, *Michigan in the War*, 267.

24. The teaching profession was near and dear to the King family. James had been an assistant teacher at the district school, and Melissa Wells had recently commented to her husband on the great success of James's sister Rebecca as an educator. Susan M. Wells to Benjamin F. Wells, June 12, 1864, Wells Papers.

25. *OR*, vol. 38, pt. 1, p. 561–62, 578; Robertson, *Michigan in the War*, 320–21; Johnson, *Body of Brave Men*, 511.

26. The dead included Private Byron Liddle, the man who had fired James's rifle and tended his wound on Missionary Ridge. *Record of Service of Michigan Volunteers*, 11:59.

27. Stoughton had married Olive, daughter of David Page of Sturgis, Michigan, in the 1850s. 1850 United States Census, Schedule 1, Sturgis Township, St. Joseph County, Mich., dwelling 99, family 99, July 29, 1850; 1860 United States Census, Schedule 1, Sturgis Village, St. Joseph County, Mich., dwelling 1852, family 1848, Aug. 1860.

28. A false rumor.

29. Ohio native William Tecumseh Sherman graduated sixth in the West Point class of 1840. He tried his hand in civilian life as a banker and lawyer before accepting the superintendence of the Louisiana State Seminary. Despite Southern sympathies, he sided with the Union and led a brigade at Bull Run before being sent to Kentucky, where feuds with the press and disagreements with the administration threatened to end his military career. He recovered to lead a division at Shiloh, earning promotion to major general, and continued under Grant in the Vicksburg Campaign and at Chattanooga, forming a close working relationship with his superior. Sherman took Grant's place atop the Military Division of the Mississippi when Grant headed to Virginia, and the two generals collaborated to coordinate their 1864 campaigns against Lee and Johnston. For more, see Marszalek, *Sherman*.

30. Jubal Early's Confederates had penetrated the Shenandoah Valley and crossed the Potomac River on July 6, directly threatening Washington. A mishmash of Federals, mostly hundred-day men, fought a hopeless battle under Lew Wallace at Monocacy Junction, buying the capital time to bolster its defenses. On July 11 the Confederates were drawn up little more than five miles from the White House, but Grant's 6th Corps, rushed back from Virginia, had arrived to man Washington's fortifications, and Early withdrew. The threat to the Union capital was a political calamity, both at home and abroad. Grant by this time had driven Lee back all the way to Petersburg in a series of turning movements, but at a much steeper human cost than Sherman incurred in Georgia, and the fighting in Virginia had degenerated into trench warfare. For more on Early's campaign, see Leepson, *Desperate Engagement*; for more on Grant and Petersburg, see Trudeau, *The Last Citadel*.

31. Darius Babcock and the 6th Michigan departed Vicksburg for St. Charles, Arkansas, the day after this letter was written. Robertson, *Michigan in the War*, 267.

32. Sherman sent Oliver Howard from the Federal left on a long march clear around the Union right. Howard reached for the railroad line south of the city in a bid to cut Hood's last line of supply. Hood responded on July 28, hurling the corps of Stephen D. Lee at the bluecoats, and the armies collided near Ezra Church. Howard was ready for trouble; the Rebels pitched headlong into breastworks. The Federal thrust against the railway was parried, but Hood's army lost close to three thousand troops while inflicting about one-fifth as many Federal casualties. Castel, *Decision in the West*, 434.

33. James W. King, "Stranger than Romance," 3, King Collection.

34. James W. King to Sarah J. Babcock, Aug. 6, 1864, King Collection; James W. King's Pension Application, Application No. 125139, Certificate No. 87706, National Archives.

35. Autopsy in the name of medical science had finally gained social acceptance in America during the war, as exemplified in the creation of the Army Medical Museum.

Previously, dissection had generally been restricted to the bodies of executed criminals and was opposed by the public to the point of triggering numerous riots; American medical students seeking to further their study were forced to choose between going abroad or robbing graves. For more, see Devine, "Whose Bodies?: Military Bodies and Control during the American Civil War," chap. 5 in *Learning from the Wounded*. James W. King, "Stranger than Romance," 4–5, King Collection.

36. James evidently discussed his hospitalization after Missionary Ridge with Jenny while he was on medical furlough. We cannot know what was said, but he did later relate two stories from his first experience as a military hospital patient:

> In [the] hospital after the battle of Mission Ridge was a comrade shot in the center of the forehead with his brain protruding. He lay there three days without anything having been done for him. He was only a few cots away from my own, and when he revived he asked how long he had lain there. He was told, and calmly said, "I will see a good many carried out [of] here before I go." He lived 34 days, and seemed to retain his reasoning faculties up to the time of his death. In that same hospital was a Scotchman by the name of McDonald, of the 19th Illinois. The ball that wounded him entered just above the right eye and came out back of his right ear. He walked from the ambulance into the hospital and the surgeon pronounced his wound a fatal one. He sat down on his cot, coolly took from his pocket a briar pipe, filled it, went to the stove and lighted it, and seemed to enjoy his smoke just as well as any man in full health. In just 23 days he went home on furlough, returned after a couple of months, and served to the end of the war.

James W. King, "Truth Is Stranger Than Fiction," 7, King Collection.

37. Actually, the regiment was still posted on the front line at this point. The 11th had been heavily engaged at the Battle of Utoy Creek on August 7, successfully seizing Confederate entrenchments after enduring a deadly charge over open field. Fifteen Wolverines were killed and fifteen wounded. Among the injured was Jenny's cousin Augustus Milo Wellman. He survived, stayed in the service, and mustered out on September 16, 1865. Johnson, *Body of Brave Men*, 522–27; Thornton, *When Gallantry Was Commonplace*, 235–37; Robertson, *Michigan in the War*, 321; *Record of Service of Michigan Volunteers*, 11:100.

38. Thornton, *When Gallantry Was Commonplace*, 238–39, 245–48; "Last Victory of the Eleventh Regiment over the Rebels," *Reporter*, Oct. 1, 1864; "The Grand Republican Rally at White Pigeon," *Reporter*, Oct. 1, 1864; James W. King, "11th Michigan Infantry," *State Republican*, Jan. 11, 1901; James W. King's Medal of Honor Nomination File, Record & Pension Office, Document Number 670252, National Archives; *Portrait and Biographical Album*, 267.

39. James's disability dilemma was faced by tens of thousands of soldiers who sought gainful employment after losing the function of their limbs. Most of these men had made their living through manual labor before the war and faced a daunting challenge securing postwar careers. *Register of Officers*, 171; Nelson, *Ruin Nation*, 200–11.

40. James W. King to Sarah J. Babcock, Apr. 10, 1865, King Collection.

41. Several Tennesseans who expressed joy over the president's fate were shot by Union soldiers. James W. King to Sarah J. Babcock, Apr. 16, 1865, King Collection; Harrell, *When the Bells Tolled*, 50–51.

42. James W. King to Sarah J. Babcock, July 30, 1865, King Collection.

43. Trowbridge was pastor of the Three Rivers First Baptist Church. There seems to have been a bit of secrecy surrounding the occasion initially; James's letters from this period suggest that the couple was concerned with how their marriage would be publicly perceived, given that the newlyweds were soon to part. *Reporter*, Mar. 7, 1863, p. 2.

44. James W. King to Sarah J. Babcock, Feb. 4, 1865, King Collection.

7. KING COTTON

1. As ex-soldiers with farming backgrounds, James and Linus were anything but typical among those who headed south to raise cotton. Securing and running a cotton plantation required a serious commitment of capital. Merely 5 percent of such individuals were farmers by trade, with approximately 90 percent consisting of businessmen and professionals. Their average age was thirty-three; Linus had recently celebrated his thirtieth birthday, but James was only twenty-three years of age. Additionally, transplanted ex-bluecoat planters were mostly officers, with nearly 40 percent having attained at least the rank of major. In one way, however, the case of these two Michiganders was not unusual. It was common for the cotton-raising veteran to choose a locale familiar from his days of military adventure, and in this respect James and Linus surely felt at home in Tennessee. Foner, *Reconstruction*, 137, 294–96, 609–12; Donald, Baker, and Holt, *Civil War and Reconstruction*, 496–98; Powell, *New Masters*, 8, 10, 13.

2. A tornado swept away the Critz house's second floor in 1921, but the building still stands and is recognized in the National Register of Historic Places. Oden, *History of Thompson's Station*, 126.

3. Federal authorities took measures to press freedmen to continue working on plantations. By choice or by coercion, many ex-slaves continued cultivating the same land they had tilled under bondage. A significant number of slaves in central Tennessee, however, had taken advantage of the lengthy Union occupation to flee their masters. Oden, *Hold Us Not Boastful*, 126; 1860 United States Census, Schedule 2, 1st district, Williamson County, Tenn., pp. 21–22, June 28, 1860; 1866 Tax Records, Civil District 4, Williamson County, Tenn.; Donald, Baker, and Holt, *Civil War and Reconstruction*, 505; Ash, *Middle Tennessee Society Transformed*, 119.

4. Judging by James's correspondence with Jenny, it appears that Squire had partnered with Melvin Mudge for the 1865 season. James W. King to Sarah J. Babcock, June 11, 1865, King Collection; James W. King to Sarah J. King, Nov. 15, 1865, King Collection.

5. New York native Linus Truman Squire was a son of Almon G. and Sally (Skeels) Squire. He had initially returned to his parents' farm in Quincy, Michigan, after mustering out of the 11th. Linus's officer's sword remains in the possession of his collateral descendants. *History of Branch County*, 440–41.

6. Probably referring to Jenny's youngest sibling, six-year-old Mary Esther Babcock. She is later referred to as Etty or Etta.

7. These payment terms were harsher than average. A typical lease agreement called for half down, with the other half due at the end of the year. As for pricing, annual lease rates for cotton land ran anywhere from three dollars per acre in parts of Alabama to twenty-two dollars per acre for prime real estate on the Mississippi River. It was unusual for field hands to be paid monthly; typically they received compensation either in wages withheld until year end or as a share of the crop. Delayed compensation was preferable for cash-strapped employers, and it discouraged the field hands from leaving in mid-season. But the offer of monthly pay may have granted James and Linus a leg up in hiring; the labor market was tight in 1866 as freedmen bided their time, holding out hope that the federal government would grant them their own land. By paying cash wages, James and Linus also maintained the option to promptly dismiss anyone who wasn't pulling their weight. Almost certainly they would have had written contracts prepared, likely with the supervision of the Freedmen's Bureau. Many Williamson County planters in this period appreciated the bureau's influence in ensuring that labor contracts were enforced. Powell, *New Masters*, 42–43; Foner, *Reconstruction*, 138–39, 171–72; Ash, *Middle Tennessee Society Transformed*, 196, 200.

8. William C. Iddings and Fiene Johnson were joined in marriage in St. Joseph County, Michigan, on February 13. James had saved their relationship in 1865; Fie had second thoughts about marriage and broke off the relationship, but she eventually reconsidered and used James as an intermediary to patch things up. Transcribed marriage record from the St. Joseph County Clerk Office, King Collection.

9. Possibly referring to Jenny's younger sister, Ruth Rozanna Babcock. The controversy involving Letitia and Rose is never explained, but later correspondence suggests that James was deeply offended by it.

10. Lemuel Pierce had married Louisa "Lizzie" Brownell on March 22. Louisa was the widow of Lemuel and Jenny's first cousin George Heman Brownell, son of Heman and Alira (Blodgett) Brownell. Louisa had two children from that marriage. As a private in the 6th Michigan Cavalry of Custer's brigade, George Brownell was wounded at Gettysburg during Custer's clash with Jeb Stuart's cavalry behind the Union lines during Pickett's Charge. Brownell's shin was broken by a shell fragment, and he received a medical furlough in late November 1863. He unsuccessfully requested a disability discharge at the expiration of his furlough and returned to duty. In May 1864 he was transferred to the Veteran Reserve Corps and died in a Louisville hospital in mid-July. In Louisa's widow's pension application, his death is variously attributed to typhoid, chronic diarrhea, dysentery, or his Gettysburg wound. Busey considers Brownell the sole Gettysburg fatality of the 6th Michigan Cavalry. George H. Brownell's Compiled Military Service File, National Archives; George H. Brownell's Pension Application, Application No. 65646, Certificate No. 49349, National Archives; Busey, *These Honored Dead*, 100.

11. Lewis G. Kies and Philip Lander, a native German, had been James's fellow clerks in Ezra Kirk's office. *Register of Officers*, 171.

12. Martha was apparently a freedwoman domestic servant, but her name does not come up again.

13. Surviving letters shed no additional light on this conflict, but perhaps it is connected with the comments Jenny's mother made in her letter of March 11, indicating that Letitia and Rose had caused Jenny and James some offense.

14. Thomas was probably Will Iddings's younger brother. 1850 United States Census, Schedule 1, Lewisburg, Union County, Pa., dwelling 1960, family 2053, Oct. 22, 1850.

15. Harriet was a freedwoman domestic servant.

16. Kate is later mentioned working as a field hand; Crockett presumably served in the same role.

17. Mr. Baur could not be identified, but subsequent letters reveal that James was trying to sell forty acres of farmland in Fabius that he had previously purchased from his father.

18. Jimmy was evidently one of the freedmen. (As we have already seen, Jenny's family also referred to James as Jimmy.) Lawson and Julia may have been children of Kate and Crockett; no other romantic relationships among the freedmen are mentioned.

19. Moderate and radical Republicans clashed over black suffrage throughout 1866. Some radicals wanted suffrage nationwide, others in the South only. Moderates were initially opposed to extending the vote for fear of the electorate's response. By spring 1867, however, many moderates, frustrated at Southern and presidential resistance to their legislative Reconstruction measures, came to favor suffrage as a means of protecting black rights in the South via the ballot box. For some, this seemed an attractive alternative to the further centralization of the government that would accompany the imposition of federal laws, or martial law, to achieve the same goal. Clearly James had become disillusioned with his freedmen employees. Northerners journeyed to Dixie anticipating that conversion to a free labor system would imbue ex-slaves with the Yankee work ethic, but these hopes were completely at odds with the highest expectation of the freedman: that freedom would bring relaxed working conditions and softer discipline. The failure of plantation hands to live up to the Northern work ethic was falsely attributed to racial inferiority by transplanted Northerners. Many Southerners had predicted that the Yankees now in their midst would ultimately reach this conclusion. James's statement was echoed among his carpetbagger peers, one of whom groused that "every Radical abolitionist at the North [should be] compelled to carry on a cotton plantation." Powell, *New Masters*, 53, 97, 106, 112–14, 117, 121–22; Donald, Baker, and Holt, *Civil War and Reconstruction*, 506, 541, 562–63.

20. "Widder" meaning "widow"; making light of his separation from Jenny.

21. The productivity and quality of work provided by freedmen employed as domestic servants was often viewed in even harsher terms than the labor of field hands. One Northern transplant wrote, "The one grand and everlasting source of misery to every Yankee resident of the South, must be the entire and overwhelming destruction of all ideas of *order in the house*." Powell, *New Masters*, 111.

22. Dio was a nickname for Jenny's brother, Darius Jr.

23. B. F. Shields owned 106 acres of land nearby. Another possible identity for this individual is William R. Shields, a saddler who lived across the county border in the direction of Spring Hill. 1866 Tax Records, Civil District 4, Williamson County, Tenn.; 1860 United States Census, Schedule 1, District No. 22, Maury County, Tenn., p. 150, dwelling 1061, family 1061, Aug. 14, 1860.

24. Likely candidates for Mr. Shaw include B. W. Shaw, who possessed 266 acres of land in the vicinity, and John A., who owned 164 acres. 1866 Tax Records, Civil District 4, Williamson County, Tenn.

25. The world was in the throes of its fourth cholera pandemic. Fatalities in Nashville peaked in the latter half of September, with seventy residents perishing on September 24 alone. Nashville had suffered prior outbreaks in 1833, 1835, 1849, 1850, and 1854, but the death toll of 1866 nearly equaled the fatalities from all of the city's previous outbreaks combined. Public medical response, however, was well organized at the federal level and informed by lessons learned during the war: see Devine, "Cholera and the Civil War: Medical Model in the Postwar Period," chap. 6 in *Learning from the Wounded*. James mentions the St. Cloud Hotel; it was erected at the corner of Church and Fifth shortly before the war. George H. Thomas lodged there before the Battle of Nashville. The general reportedly made a point of checking out before the battle, just in case he did not return to settle his bill. Bowling, *Cholera in Nashville*, 13–14; Hoobler, *Guide to Historic Nashville*, 75.

26. James refers to his and Jenny's unborn child as J. Webster. The Kings would later name a son after Daniel Webster Holbrook.

27. Andrew Johnson had just completed his calamitous speaking tour, the "Swing Around the Circle." Against all precedent, Johnson personally campaigned against Republican candidates for the upcoming congressional elections and responded to hecklers with vitriol that embarrassed his supporters and enraged his opponents. During one of the president's speeches, a jeerer called for the hanging of Jefferson Davis, and Johnson responded by recommending the same treatment for Republican congressmen Thaddeus Stevens and Wendell Phillips. The president's demeanor further weakened his support in the North and helped set the stage for a landslide Republican victory in the 1866 elections. Regarding land confiscation, in February Thaddeus Stevens had pressed to redistribute ex-Rebels' estates to freedmen, but Congress showed little inclination in that direction. Critz's three sons (Zachariah, John M., and Thomas Leanard) had served together in the Confederate army. Foner, *Reconstruction*, 245–46, 264–65; Oden, *Hold Us Not Boastful*, 126.

28. Mark B. Price was one of the clerks in Kirk's office. *Register of Officers*, 170.

29. The first quote ("Happy they …") is from James Thomson's "Spring." The second ("How blest …") is from a wedding hymn by Anna Laetitia Barbauld.

30. Josh may have been one of the field hands. He is not mentioned again.

31. James Preston Rexford was married to Ann Adelia Carmer. He practiced law in Detroit until 1862, when he joined the 4th Michigan Cavalry. In February 1863 Lieutenant Rexford impressed his superiors by leading a bold saber charge against mounted Confederates near Murfreesboro. In 1865 he took up legal practice in Nashville. Rexford later helped found a college for freedmen in Nashville, worked as an IRS collector, and became editor of the *Nashville Bulletin* shortly before his death on August 3, 1873. His brother, William Henry Rexford, also an attorney, served as a captain in the 24th Michigan Infantry and was wounded in that regiment's famous stand at Gettysburg on July 1, 1863. *Historical Catalogue of Students*, 7; Moore, *History of Michigan*, 3:1277; Robertson, *Michigan in the War*, 645, 914; Rexford, *Genealogical History*, 20–21.

32. After the war, many freedmen preferred their own spiritual gatherings over attending white churches, which typically offered a less animated style of worship. Ash, *Middle Tennessee Society Transformed*, 220.

33. It is later mentioned that Aunty performed housework duties.

34. The poetry excerpts are from John Finley's "Bachelor's Hall."

35. "Having the mully grubs" is slang for being sulky.

36. James's father had offered to sell him forty acres of the family farmland in Fabius.

37. Robert was one of the field hands.

38. It seems likely that Thomas Iddings was also renting land from Critz—James and Linus were renting about two-thirds of the Critzes' acreage.

39. Eliza King's nephew Edwin J. Dickinson had fought in Arkansas as a first lieutenant in Merrill's Horse (the 2nd Missouri Cavalry). *Record of Service of Michigan Volunteers*, 45:1, 45:13.

40. The Civil Rights Act of 1866 and the Freedman's Bureau Bill constituted the moderate Republican response to ongoing discrimination against freedmen in the South, which was most notably expressed in the formulation of the Black Codes (harsh laws passed in some Southern states to curtail freedmen's rights). The Civil Rights Act raised blacks toward full equality before the law. Johnson's veto of these two bills ended any remaining hope for cooperation between Congress and the president. The Civil Rights veto was overridden by Congress in April, marking a shift from presidential to congressional dominance of Reconstruction legislation. It is curious that James opposed legislation central to the Republican platform. Perhaps he opposed further expansion of the federal government, but it seems more likely that this was a short-term reaction to disappointment in his freedmen employees. James mentions impeachment; radical Republicans would indeed push for exactly that when Congress reconvened, citing Johnson's efforts to obstruct congressional legislation. For the time being, moderate Republicans deterred radical efforts in this direction for fear that impeachment would trigger voter backlash. Foner, *Reconstruction*, 198–201, 242–51; Donald, Baker, and Holt, *Civil War and Reconstruction*, 528–35, 567.

41. Critz leased 200 acres to James and Linus but apparently continued to work some or all of the balance of his land, about a hundred acres. As speculated in a prior note, he may also have leased some land to Thomas Iddings.

42. William Isaac Kinkade lived nearby but did not own land. He was a veteran of the Union 25th Kentucky Infantry. James mentions his wife, Nancy, in a subsequent letter, wherein it becomes apparent that the Kinkades were also leasing cotton land nearby. 1866 Tax Records, Civil District 4, Williamson County, Tenn.

43. William Lewis Stoughton was brevetted major general on March 13, 1865. He resumed legal practice that year and secured election to attorney general of Michigan in 1866. In 1867 he served as a delegate for St. Joseph County at Michigan's constitutional convention. The 11th Michigan's beloved colonel was then elected to the U.S. House of Representatives in 1868 and reelected in 1870. *Annual Report Attorney General 1883*, 79; Lord and Brown, *Debates and Proceedings*, 1:1.

44. Manda and Melvina were freedwomen. Presumably Kate and Crockett were expecting.

45. James, however, was not the only Northerner in Dixie sleeping with a gun under his pillow. Sectional animosities heated up as bitter disputes between President Johnson

and radical Republicans preceded the 1866 elections. Some of Johnson's political gaffes have already been mentioned. He further amplified the swelling hostilities by sacking more than sixteen hundred Republican postmasters. Republicans in turn stepped up their anti-Southern rhetoric in an effort to secure the vote of the newly formed Grand Army of the Republic. Trapped in an increasingly tense and confrontational atmosphere, many Yankees in the South came to fear for their personal safety. Tennessee had already been the scene of significant violence in May, when deadly race riots broke out in Memphis. Powell, *New Masters*, 132; Donald, Baker, and Holt, *Civil War and Reconstruction*, 551–55.

46. For more about Ewell, see Pfanz, *Richard S. Ewell*.

47. Ewell lost his left leg to amputation at the Battle of Groveton in 1862. General Harlan cannot be identified with certainty. However, Ewell was acquainted with local farmer Benjamin Harlan, a first cousin of John Marshall Harlan, former adjutant general and attorney general of Kentucky. John Marshall Harlan had been referred to as "general" ever since becoming adjutant general in 1851; possibly he was Ewell's companion. The Harlans were prominent slaveholders, but John Marshall Harlan opposed secession and became colonel of the Union 10th Kentucky Infantry. He would serve on the U.S. Supreme Court from 1877 to 1911 and gain a reputation for his powerfully stated dissenting opinions—most notably in Plessy v. Ferguson—championing racial equality. For more, see Beth, *John Marshall Harlan*. Benjamin Harlan to Richard S. Ewell, Nov. 15, 1866, Brown-Ewell Papers, 1803–1919, Tennessee State Library and Archives.

48. James and Linus were making their way north. George W. Trimble owned 200 acres of land in that direction; Dr. John W. Steel, 95 acres; John Buchanan Ridley, 377 acres. Ridley was master to a dozen slaves before the war. His daughter Julia married Jacob Critz's son Thomas in July 1865. *Map of Williamson County*; 1861 and 1866 Tax Records, Civil District 4, Williamson County, Tenn.; Bowman, *Historic Williamson County*, 164; Oden, *Hold Us Not Boastful*, 126.

49. Pope's Chapel was constructed on land donated by Methodist minister John Pope in 1818. The structure was demolished by a tornado in 1910. The preacher James mentions is probably Virginia native Henry B. North. Brandt, *Touring the Middle Tennessee Backroads*, 126–27; 1860 United States Census, Schedule 1, District 1, Williamson County, Tenn., p. 63, dwelling 488, family 468, June 22, 1860.

50. A local history mentions that a Colonel Helm resided at the intersection of Evergreen and Pope's Chapel sometime during the nineteenth century. Helm does not seem to have served either side in the Civil War. In 1866 local landowners sharing the Helm surname included Fielding and Henderson Helm. Bowman, *Historic Williamson County*, 164; 1866 Tax Records, Civil District 4, Williamson County, Tenn.

51. Another reference to James's attempts to find a buyer for his farmland in Fabius. Vet could not be identified.

52. Freedmen were quick to recognize the power of collective bargaining, and such court disputes became commonplace. One Southern justice of the peace declared that at least three quarters of the cases coming before him in this period involved ex-slaves violating their labor contracts. Powell, *New Masters*, 104, 119.

53. William Isaac Kinkade and his wife Nancy were evidently leasing a plantation; 1866 tax records show that they were not local landowners. They appear in the 1870 census in Covington, Kentucky, Nancy's native state. William was born in Tennessee.

1866 Tax Records, Civil District 4, Williamson County, Tenn.; 1870 United States Census, Schedule 1, Covington, Warren County, Ky., p. 29, dwelling 184, family 189, Oct. 11, 1870.

54. James S. Porter was another clerk in Kirk's office. *Register of Officers*, 171.

55. With county courthouses generally under the control of native whites, the rulings in interracial cases did not often favor the freedmen, except in instances of intervention by the Freedmen's Bureau. Ash, *Middle Tennessee Society Transformed*, 198, 220.

56. James had begun studying shorthand during the siege of Chattanooga. His efforts in this area are discussed in greater detail in the next chapter.

57. The identity of the Alabama plantation owner could not be determined with certainty, but there are two likely candidates. James later mentions that the plantation was at the foot of a mountain. The owner may have been one of the sons of prominent plantation owner Jonathan Burleson, who owned a great deal of land—and before the war, a hundred slaves—on and around Burleson Mountain. Burleson came to Alabama with his family in 1808, fought under Andrew Jackson during the War of 1812, married Elizabeth Byrd, and settled south of Decatur in 1818. The couple had thirteen children. Burleson served as justice of the peace and county commissioner. He died in 1866, but two of his sons, Aaron and Dabney, still resided in Decatur. Another good candidate is Jonathan Ford, a Unionist who owned a plantation near the present location of North Alabama Regional Hospital. Ford was Morgan County's delegate to Alabama's secession convention in January 1861, where his nay vote made no secret of his loyalties. Ford was appointed commissioner of oaths after the war, openly supported Republicans running for office, and was elected probate judge in 1868. James W. King, "11th Michigan Infantry," *State Republican*, Jan. 11, 1901; Knox, *History of Morgan County*, 50, 101–3, 105–6, 181, 201, 205; *Northern Alabama*, 65; Jenkins and Knox, *Story of Decatur, Alabama*, 105–6.

58. Jenkins and Knox, *Story of Decatur, Alabama*, 124–25, 240–41.

59. Pests, faulty seeds, and poorly timed rains plagued plantations even as the bottom fell out of the cotton market. Prices plummeted to a paltry fourteen cents per pound—down almost 95 percent from the lofty market peak of 1865. On the legislative side, freedmen were enfranchised at the same time many ex-Confederate political and military leaders had their voting and office-holding rights revoked. Additionally, with the exception of Tennessee, the ex-Confederate states were divided into five military districts. Sectional differences were enflamed and became further exacerbated under financial stress. Powell, *New Masters*, 123, 132, 145–46, 148–51; Donald, Baker, and Holt, *Civil War and Reconstruction*, 558–59.

60. King, "Personal Recollections," 179.

61. James W. King, "11th Michigan Infantry," *State Republican*, Jan. 11, 1901.

62. The Klan, founded in Tennessee, expanded into northern Alabama by 1867. By November, at least one Republican living south of Decatur was receiving KKK warnings—one of which originated from Blount County's probate judge. By the summer of 1868, Klan members would unabashedly walk the streets of Decatur, murder freedmen by night, and drive off Northerners and scalawags residing in the area. The Klan acquired such notoriety that even such acts carried out by unassociated individuals tended to be ascribed to the Klan by default. "Opinions of James W. King," *News Reporter*, Oct. 15, 1903; James W. King to Sarah J. King, Jan. 8, 1868, King Collection; Trelease, *White Terror*, 81–82, 87; Storey, *Loyalty and Loss*, 218.

63. 1860 United States Census, Schedule 1, 7th Ward, Cincinnati, Hamilton County, Ohio, dwelling 632, family 1494, July 20, 1860; 1870 United States Census, Schedule 1, Subdivision No. 39, Morgan County, Ala., dwelling 120, family 119, June 4, 1870; Hurd, "Battle of Collierville," 5:252–53.

64. Daniel Webster Holbrook's younger brother, George, was a clerk in Kirk's office. James W. King to Sarah J. King, Oct. 29, 1865, King Collection; 1850 United States Census, Schedule 1, Coldwater, Branch County, Mich., dwelling 36, family 45, July 22, 1850; *Register of Officers*, 171.

65. Possibly there was some interplay between James's financial woes and the Klan's actions against him, but there is no further evidence regarding this. James W. King to Sarah J. King, Jan. 8, 1868, King Collection.

8. The Battle of Life

1. James's postwar well-being hinged on the fact that his disabled limb could still grasp a pen—and wield it with unusual expertise. Many veterans, particularly amputees, were not so lucky, and never regained the ability to support themselves. King, "Personal Recollections," 179–80; Nelson, *Ruin Nation*, 200–11.

2. James W. King to Sarah J. King, Jan. 28, Apr. 19, 29, and May 18, 1868, King Collection.

3. James W. King's Pension Application, Application No. 125139, Certificate No. 87706, National Archives; James W. King to Sarah J. King, Apr. 19, May 18, and June 3, 1868, King Collection; 1870 United States Census, Schedule 1, 3rd Ward, City of Lansing, Ingham County, Mich., pp. 21–22, dwelling 183, family 187, July 13, 1870; *Portrait and Biographical Album*, 269.

4. King, "Personal Recollections," 179; James W. King to Sarah J. King, Feb. 5, 14, 22, 28, Apr. 15, and May 16, 1869, King Collection.

5. James W. King to Sarah J. King, Apr. 18 and May 16, 1869, King Collection; Washington, *Records of Field Offices*, 8, 10; "Local News Brevities: At the Capital," *Lansing (Mich.) Tri-Weekly Republican* (hereafter cited as *Republican*), July 2, 1881.

6. James W. King to Sarah J. King, June 6, 1869, King Collection; James W. King, "Good-By and God Speed," *Republican*, Jan. 2, 1886, photocopy in the King Collection; "James W. King," 2–3; *Proceedings of Michigan Press*, 11.

7. James W. King, "Good-By and God Speed," *Republican*, Jan. 2, 1886; "James W. King," 2–3.

8. Durant, *History of Ingham*, 159–60; *Lansing and Its Yesterdays*, 32. The manual was specified in the legislation as being

for the use of members and officers of both Houses in this and the next Legislature, and the State officers,—said manual to contain the Constitutions of the United States and of this State, with all amendments thereto; the rules and joint rules of the Senate and House of Representatives of this State; a diagram of the Senate Chamber and Representative Hall; names, ages, occupation, and residence of members of both Houses; a map showing the Senatorial and Congressional districts, the judicial circuits of the State, the various Senatorial and Representative districts of the State, with the population thereof; the votes for President in eighteen hundred and seventy-two and eighteen hundred and sixty-eight; the postoffices, newspapers, banking institutions, railroad routes, a history of the Constitutional conventions, and the latest statistics of the educational, charitable, reformatory, and penal institutions, the table of equalization for eighteen hundred and seventy-one, and such other statistical matter as is usually contained in the work; the same to be printed and bound in the usual style by the State printer, and the compiler of said manual to receive for his services such sum as shall be fair and adequate.

General Acts, 576–77.

9. James W. King, "Reunion of Michigan Veterans," unidentified newspaper, Aug. 25, 1873; "Veteran Reunion at Colon," *Tribune*, Aug. 25, 1891, photocopy in the King Collection; "Re-Union of the 11th. Mich.," *Mendon (Mich.) Weekly Globe*, Aug. 25, 1893, photocopy in the King Collection.

10. *Lansing and Its Yesterdays*, 32; Durant, *History of Ingham*, 159–60; *Proceedings of Michigan Press*, 9; "Opinions of James W. King," *News Reporter*, Oct. 15, 1903.

11. *Proceedings of Michigan Press*, 10–12; James W. King, "Good-By and God Speed," *Republican*, Jan. 2, 1886; O. L. Spaulding to James W. King, Apr. 15, 1882, King Collection; B. M. Cutcheon to James W. King, Mar. 21, 1884, King Collection.

12. *Republican*, Jan. 29, 1880, p. 2; *Republican*, Nov. 25, 1880, p. 2; "Editorial Notes," *Republican*, Jan. 4, 1881; *Republican*, Apr. 23, 1881, p. 2; "Sojourner Truth in Lansing," *Republican*, June 4, 1881; "Entertaining Miscellany: Women of Worth," *Republican*, Mar. 7, 1882; *Republican*, Nov. 19, 1884, p. 2; *Republican*, Apr. 22, 1885, p. 2.

13. *Republican*, Feb. 11, 1885, p. 2; "The Temperance Question," *Republican*, Sept. 23, 1885; *Republican*, Jan. 27, 1880, p. 2; "The Tariffs: Shall It Be for Revenue or Protection?" *Republican*, Apr. 5, 1881; *Republican*, Apr. 21, 1881, p. 2; *Republican*, Apr. 1, 1885, p. 2.

14. "Local News Brevities: At the Capital," *Republican*, Jan. 4, 22, Apr. 30, and July 2, 1881, Mar. 25, 1882; "Death of Captain E. G. Hall," *Republican*, Apr. 14, 1881.

15. *Portrait and Biographical Album*, 268–69.

16. *Lansing City and Ingham*, 1:36.

17. The new *Republican* owners were Thorp & Godfrey, who converted it into a daily. The paper would ultimately become the *Lansing State Journal*. "Opinions of James W. King," *News Reporter*, Oct. 15, 1903; Cowles, *Past and Present Lansing*, 113.

18. James W. King, "Good-By and God Speed," *Republican*, Jan. 2, 1886.

19. King, "Another Pioneer Gone," 10:183–84; "St. Joseph County: Proceedings," 5:504; *Portrait Record of Kalamazoo*, 646.

20. King, "Another Pioneer Gone," 10:183–84; 1880 United States Census, Schedule 1, Battle Creek, Calhoun County, Mich., p. 11, family 10, dwelling 10, June 10, 1880.

21. James W. King, stenographer appointment for Fifteenth Judicial Circuit, State of Michigan, Oct. 7, 1887, King Collection.

22. An extract of James's speech, in his handwriting, is preserved in the Louisiana Research Collection at Tulane University, filed with the Louisiana Historical Association Collection. It is written on the letterhead of the Chickamauga, Chattanooga and Missionary Ridge Military Park Commission of Michigan. The entire speech was published in Belknap's work as cited below. For more about the creation of the military

park, see Smith, *Chickamauga Memorial.* "Re-Union of the 11th. Mich.," *Mendon (Mich.) Weekly Globe,* Aug. 25, 1893; Belknap, *Michigan Organizations,* 241–50; "The Chickamauga Commission," *Tribune,* Nov. 3, 1893, photocopy in the King Collection.

23. Eva F. King, Certified Record of Death, Centreville, St. Joseph County, Mich., May 21, 1895, certified copy.

24. Heart disease was far more prevalent among war veterans than among the general population. James W. King's Pension Application, Application No. 125139, Certificate No. 87706, National Archives; Marten, *Sing Not War,* 86.

25. Adams's sword has since relocated to the Confederate Memorial Hall Museum, which also possesses two bullets (presumably from the rifles of the 11th Michigan) that struck General Adams. James W. King, "11th Michigan Infantry," *State Republican,* Jan. 11, 1901.

26. John W. King to James W. King, July 18, 1898, King Collection; Stilwell, *33d Michigan Volunteer Infantry,* 81–83.

27. Hill, "National Stenographer's Association," 78; Howard, *Phonographic Magazine,* 17:312.

28. 1900 United States Census, Schedule 1, Fabius, St Joseph County, Mich., p. 4, dwelling 69, family 69, June 6–7, 1900; 1900 United States Census, Schedule 1, Fabius, St Joseph County, Mich., p. 9, dwelling 197, family 198, June 17–18, 1900; "James W. King: A Sketch of the Life of This Prominent and Well Known Man," *News Reporter,* Oct. 15, 1903, photocopy in the King Collection; Howard, *Phonographic Magazine,* 14:94; James W. King's Pension Application, Application No. 801326, Certificate No. 651535, National Archives.

29. James W. King, "11th Michigan Infantry," *State Republican,* Jan. 11, 1901; King, "Eleventh Michigan Infantry," King Collection.

30. Robertson, *Michigan in the War,* 319; Belknap, *Michigan Organizations,* 121.

31. "Letter from the Eleventh Mich.," *Reporter,* Dec. 12, 1863; "Another Letter from the Eleventh Regiment," *Reporter,* Dec. 12, 1863; *Medal of Honor,* 15; Broadwater, *Medal of Honor Recipients,* 315–20; Millbrook, *Study in Valor,* 118–19; Beyer and Keydel, *Deeds of Valor,* 1:269–70.

32. James W. King's Medal of Honor Nomination File, Record and Pension Office, Document No. 670252, National Archives; Millbrook, *Study in Valor,* 72, 75, 95, 98; *Medal of Honor,* 16.

33. Linus Squire had remarried, to Susan Johnson Snyder. The husband and wife attended school and obtained medical degrees together, but Linus spent the balance of his life in Washington, D.C., working for the federal government in the pension and post offices. *History of Branch County,* 441.

34. In at least one case, a soldier initially rejected based on the March 1903 active duty requirement successfully obtained a medal by appealing after the April 1904 legislation passed. James W. King's Medal of Honor Nomination File, Record and Pension Office, Document No. 670252, National Archives; Broadwater, *Medal of Honor Recipients,* 318–20; *Medal of Honor,* 16–17; Millbrook, *Study in Valor,* 123.

35. James W. King's Pension Application, Application No. 125139, Certificate No. 87706, National Archives.

36. "James W. King: A Sketch of the Life of This Prominent and Well Known Man," *News Reporter*, Oct. 15, 1903.

37. Mrs. B. E. Andrews was almost certainly Lucy, the wife of Bishop E. Andrews, a veteran of the 25th Michigan Infantry. Lucy and Bishop are recorded as local lawyers in the 1900 census, which lists them as close neighbors to John Willard King and Charles Rice. "James W. King: A Sketch of the Life of This Prominent and Well Known Man," *News Reporter*, Oct. 15, 1903; "Opinions of James W. King," *News Reporter*, Oct. 15, 1903; 1900 United States Census, Schedule 1, Lockport Township, St. Joseph County, Mich., supervisor's district 4, enumeration district 117, p. 3, dwelling 53, family 54, June 4, 1900; *Record of Service of Michigan Volunteers*, 25:5.

38. "James W. King: A Sketch of the Life of This Prominent and Well Known Man," *News Reporter*, Oct. 15, 1903; "Opinions of James W. King," *News Reporter*, Oct. 15, 1903.

39. 1880 United States Census, Schedule 1, 3rd Ward City of Lansing, Ingham County, Mich., p. 4, dwelling 33, family 36, 2 June 1880; 1880 United States Census, Schedule 1, 3rd Ward City of Lansing, Ingham County, Mich., p. 4, dwelling 35, family 38, June 2, 1880.

40. This letter was reprinted in the *Three Rivers News Reporter*. Henry Humphrey, "Tribute to Mr. King," *News Reporter*, Oct. 15, 1903, photocopy in the King Collection.

41. Melvin Mudge to James W. King, Feb. 9, 1901, King Collection; James W. King's Pension Application, Application No. 801326, Certificate No. 651535, National Archives.

APPENDIX A

1. Broadwater, *Medal of Honor Recipients*, 322.

2. Beyer and Keydel, *Deeds of Valor*, 1:7, 1:206, 1:289, 1:313–15; *Ninth Reunion*, 27–29, 42–44; *Journal of the Nineteenth*, 22.

3. Ironically, the board of officers that tightened the medal standards was headed by Arthur MacArthur Jr., winner of a belated (1890) Medal of Honor for the Battle of Missionary Ridge. He was the father of General Douglas MacArthur, who would also receive the medal, for service in World War II. Millbrook, *Study in Valor*, 46–47, 123; Broadwater, *Medal of Honor Recipients*, 320; James W. King's Medal of Honor Nomination File, Record and Pension Office, Document No. 670252, National Archives.

4. James W. King's Medal of Honor Nomination File, Record and Pension Office, Document No. 670252, National Archives; Millbrook, *Study in Valor*, 90, 110, 123, 146; *Medal of Honor*, 5, 15, 432; Broadwater, *Medal of Honor Recipients*, 318–20.

APPENDIX B

1. James Moody and his guide, who would capture Montanye, placed these events on March 15, but Montanye was bearing a letter from Washington dated March 27, and the *New York Gazette* placed his trek through Ramapo Pass more believably on March 29, with his arrival in New York occurring on April 1. Note that the spelling of the surname Montanye has varied over time (e.g. Montaigne or Montagne). Shenstone, *So Obstinately Loyal*, 111–13; Lossing, *Field-Book of the Revolution*, 1:781, 1:781n2; Moody, *Lieut. James Moody's Narrative*, 36; "Mail from New-York," *London Political Magazine*, 343–44.

2. Montanye's capture was glorified—with little concern for accuracy—in nineteenth century historical fiction, most notably in Tomlinson's *Stories of the American Revolution* and in the 1856 *United States Magazine* article, "The Intercepted Messenger of Ramapo Pass," by Elizabeth Oakes Smith. Shenstone, *So Obstinately Loyal*, 110–13, 139; Moody, *Lieut. James Moody's Narrative*, 36–37, 56; Sir Henry Clinton to Lord George Germain, Apr. 5, 1781, in Stevens, *Campaign in Virginia 1781*, 1:384; "Mail from New-York," *London Political Magazine*, 343–44.

3. Shenstone, *So Obstinately Loyal*, 112–13; Lossing, *Field-Book of the Revolution*, 1:781, 1:781n2; Coffin, *Boys of '76*, 382; Pennypacker, *George Washington's Spies*, 212–14; Casey, "Remarks of William J. Casey," 10–13; Knott, *Secret and Sanctioned*, 24–26; Quinlan, *History of Sullivan County*, 448; *Portrait and Biographical Album*, 266; Application for Membership, Daughters of the American Revolution, National No. 141450, DAR Library, Washington, D.C.

4. George Washington to Noah Webster, July 31, 1788, in Webster, *Collection of Papers*, 166–67.

5. Some historians, including Lossing, have placed Montanye's capture immediately prior to Washington's departure for Virginia in August and suggested that Montanye directly enabled the American victory at Yorktown. But contemporaneous sources place his capture in March, and there is no evidence that he was recaptured in August. "Mail from New-York," *London Political Magazine*, 343–44; Sir Henry Clinton to Earl Cornwallis, June 8, 1781, in Stevens, *Campaign in Virginia 1781*, 2:15; Lossing, *Field-Book of the Revolution*, 1:781, 1:781n2; Coffin, *Boys of '76*, 382; Pennypacker, *George Washington's Spies*, 212–14; Casey, "Remarks of William J. Casey," 10–13; Knott, *Secret and Sanctioned*, 24–26; Quinlan, *History of Sullivan County*, 448; *Portrait and Biographical Album*, 266.

6. The elder Samuel King served in James McClaughrey's Second Regiment of the Ulster County militia. He may have seen action at Fort Montgomery and at Minisink. Riker, *Revised History of Harlem*, 790–91; Application for Membership, Daughters of the American Revolution, National No. 410890, DAR Library, Washington, D.C.; Silliman, *St. Joseph in Homespun*, 130; Roberts, *New York in Revolution*, 209; Clearwater, *History of Ulster County*, 157–58.

7. Silliman, *St. Joseph in Homespun*, 129; Application for Membership, The Empire State Society of the Sons of the American Revolution, National No. 11982, accessed via Ancestry.com; James W. King to Sarah J. King, Jan. 8, 1868, King Collection.

Bibliography

ARCHIVAL COLLECTIONS

Archives of Michigan, Lansing
>   Records of the Michigan Military Establishment. Record Group 59–14.

Bentley Historical Library, University of Michigan, Ann Arbor
>   Bordner, Benjamin F., Papers, 1862–1864
>   Fox, William H., Letters, 1862–1863
>   Martin, James. Letters, 1861–1864
>   Raymond, William E., Papers, 1863–1864
>   Van Buren, Anson De Puy. Papers, 1846–1885
>   Wells Family Papers, 1857–1902
>   White, Aaron B., Papers, 1863–1864

Daughters of the American Revolution DAR Library, Washington, D.C.
>   Application Records

National Archives and Records Administration, Washington, D.C.
>   Records of the Adjutant General's Office, 1780s–1917. Record Group 94
>   Records of the Department of Veterans Affairs. Record Group 15

Tennessee State Library and Archives, Nashville
>   Brown-Ewell Papers, 1803–1919

Western Michigan University Archives and Regional History Collections, Kalamazoo
>   Dudd, John. Collection, 1958
>   King, James W., Collection, 1861–1903
>   Masterson, Carroll. Collection, 1861–1865

NEWSPAPERS

Constantine (Mich.) Weekly Mercury and St. Joseph County Advertiser
Frank Leslie's Illustrated Newspaper
Lansing (Mich.) State Republican
Lansing (Mich.) Tri-Weekly Republican
Mendon (Mich.) Weekly Globe
New York Times
Three Rivers (Mich.) News Reporter
Three Rivers (Mich.) Reporter
Three Rivers (Mich.) Tribune
Three Rivers (Mich.) Western Chronicle

PUBLISHED PRIMARY SOURCES

Annual Report of the Adjutant General of the State of Michigan for the Year 1862: Together with a Supplementary Report. . . . Lansing, Mich.: John A. Kerr, 1863.
Annual Report of the Adjutant General of the State of Michigan for the Year 1863. Lansing, Mich.: John A. Kerr, 1864.
Annual Report of the Adjutant General of the State of Michigan for the Year 1864. Lansing, Mich.: John A. Kerr, 1865.
Annual Report of the Attorney General of the State of Michigan for the Year 1883. Lansing, Mich.: W. S. George, 1884.
Baillie, Joanna, ed. *Collection of Poems, Chiefly Manuscript, and from Living Authors.* London: Longman, Hurst, Rees, Orme, and Brown, 1823.
"Capt. E. G. Hall." *Student's Journal* 10, no. 5 (May 1881): 3–4.
Edmonds, Sarah Emma. *Memoirs of a Soldier, Nurse, and Spy: A Woman's Adventures in the Union Army.* Dekalb: Northern Illinois Univ. Press, 1999.
Ely, Ralph. *The Diary of Captain Ralph Ely of the Eighth Michigan Infantry: With the Wandering Regiment.* Edited by George M. Blackburn. Mount Pleasant: Central Michigan Univ. Press, 1965.
General Acts and Joint and Concurrent Resolutions of the Legislature of the State of Michigan Passed at the Regular Session of 1873, with an Appendix. Lansing, Mich.: W. S. George, 1873.
Gillaspie, Ira. *The Diary of Ira Gillaspie of the Eleventh Michigan Infantry.* Edited by Daniel B. Weber. Mount Pleasant: Central Michigan Univ. Press, 1965.
Hewett, Janet B., et al., eds. *Supplement to the Official Records of the Union and Confederate Armies.* Vol. 6, Part 1. Wilmington, N.C.: Broadfoot, 1996.
Hicks, Borden M. "Personal Recollections of the War of the Rebellion." In *Glimpses of the Nation's Struggle, Sixth Series: Papers Read before the Minnesota Commandery of the Military Order of the Loyal Legion of the United States, January 1903–1908,* 519–44. Minneapolis: Aug. Davis, 1909.
Hurd, Ethan O. "The Battle of Collierville." In *Sketches of War History, 1861–1865,* 5:243–54. Cincinnati: Robert Clarke, 1903.

Journal of the Nineteenth Annual Session of the National Encampment, Grand Army of the Republic, Portland, Maine, June 24th and 25th, 1885. Toledo, Ohio: Montgomery and Vrooman, 1885.

Kimball, William Horton. *Among the Enemy: A Michigan Soldier's Civil War Journal.* Edited by Mark Hoffman. Detroit: Wayne State Univ. Press, 2013.

King, James W. "Another Pioneer Gone—Benjamin M. King." In *Pioneer Collections: Collections and Researches Made by the Pioneer Society of the State of Michigan,* 2nd ed., 10:183–84. Lansing, Mich.: Wynkoop Hallenbeck Crawford, 1908.

———. "Personal Recollections." *Journal of Commercial Education* 9 (1894): 179–80.

Lansing City and Ingham County Directory, 1883–1884. . . . Vol. 1. Detroit: R.L. Polk, 1883.

Lord, William Blair, and David Wolfe Brown. *The Debates and Proceedings of the Constitutional Convention of the State of Michigan, Convened at the City of Lansing, Wednesday, May 15th, 1867.* Vol. 1. Lansing, Mich.: John A. Kerr, 1867.

Mayo, Perry. *The Civil War Letters of Perry Mayo.* Edited by Robert W. Hodge. East Lansing: Michigan State Univ. Press, 1967.

The Medal of Honor of the United States Army. Washington, D.C.: United States Government Printing Office, 1948.

Moody, James. *Lieut. James Moody's Narrative of His Exertions and Sufferings in the Cause of Government, Since the Year 1776.* 2nd ed. London: Richardson and Urquhart, 1783.

Ninth Reunion of the 37th Regiment O.V.V.I., St. Mary's, Ohio, Tuesday and Wednesday, September 10 and 11, 1889. Toledo, Ohio: Montgomery and Vrooman, 1890.

Official Records of the Union and Confederate Navies in the War of the Rebellion. Vol. 22. Washington, D.C.: Government Printing Office, 1908.

Paddington, John Henry. *Dear Sarah: Letters Home from a Soldier of the Iron Brigade.* Edited by Coralou Peel Lassen. Bloomington: Indiana Univ. Press, 1999.

Proceedings of the Eighth Annual Reunion of the Eleventh Michigan Infantry and Fourth Michigan Battery, Held at Centreville, Mich., August 24, 1875. Three Rivers, Mich.: W. H. Clute, [1875].

Register of Officers and Agents, Civil, Military, and Naval, in the Service of the United States, on the Thirtieth September, 1865. . . . Washington, D.C.: Government Printing Office, 1866.

Sanders, Charles W. *The School Reader: Fourth Book, Containing Instructions in the Elementary Principles of Reading, and Selected Lessons from the Most Elegant Writers. For the Use of Academies and the Higher Classes in Common and Select Schools.* New York: Mark H. Newman, 1842.

Sears, Stephen W., ed. *For Country, Cause and Leader: The Civil War Journal of Charles B. Haydon.* New York: Ticknor and Fields, 1993.

Stevens, Benjamin Franklin, ed. *The Campaign in Virginia 1781: An Exact Reprint of Six Rare Pamphlets on the Clinton-Cornwallis Controversy.* . . . 2 vols. London: Trafalgar Square, Charing Cross, 1888.

"St. Joseph County: Proceedings at the Annual Meeting of the St. Joseph County Pioneer Society, at Centreville, Mich., June 14, 1882." In *Pioneer Collections: Report of the Pioneer Society of the State of Michigan, Together with Reports of County, Town, and District Pioneer Societies,* 5:504–18. Lansing, Mich.: W. S. George, 1884.

Tourgée, Albion W. *A Fool's Errand: A Novel of the South during Reconstruction.* New York: Cosmo Classics, 2005.

Van Buren, Anson De Puy. "Deacon Isaac Mason's Early Recollections of Michigan." In *Pioneer Collections: Report of the Pioneer Society of the State of Michigan, Together with Reports of County, Town, and District Pioneer Societies*, 5:397–402. Lansing, Mich.: W. S. George, 1884.

———. *Jottings of a Year's Sojourn in the South, or, First Impressions of the Country and Its People; With a Glimpse at School-Teaching in That Southern Land, and Reminiscences of Distinguished Men*. Battle Creek, Mich.: n.p., 1859.

The War of the Rebellion: A Compilation of the Official Records of the Union and Confederate Armies. 128 vols. Washington, D.C.: Government Printing Office, 1880–1901.

Washington, Reginald. *M1911: Records of the Field Offices for the State of Tennessee, Bureau of Refugees, Freedmen, and Abandoned Lands, 1865–1872*. Edited by Benjamin Guterman. Washington, D.C.: United States Congress and National Archives and Records Administration, 2005.

Webster, Noah. *A Collection of Papers on Political, Literary and Moral Subjects*. New York: Webster and Clark, 1843.

Willcox, Orlando B. *Forgotten Valor: The Memoirs, Journals, and Civil War Letters of Orlando B. Willcox*. Edited by Robert Garth Scott. Kent, Ohio: Kent State Univ. Press, 1999.

Wilterdink, John Anthony. *My Country and Cross: The Civil War Letters of John Anthony Wilterdink, Company "I," 25th Michigan Infantry*. Edited by Albert H. McGeehan. Dallas: Taylor, 1982.

Wittenberg, Eric J., ed. *At Custer's Side: The Civil War Writings of James Harvey Kidd*. Kent, Ohio: Kent State Univ. Press, 2001.

SECONDARY SOURCES

Books

Adams, George Worthington. *Doctors in Blue: The Medical History of the Union Army in the Civil War*. New York: Henry Schuman, 1952.

American Biographical History of Eminent and Self-Made Men, with Portrait Illustrations on Steel, Michigan Volume. Cincinnati: Western Biographical, 1878.

Armstrong, Zella. *The History of Hamilton County and Chattanooga, Tennessee*. Vol. 2. Johnson City, Tenn.: Overmountain, 1993.

Ash, Stephen V. *Middle Tennessee Society Transformed, 1860–1870: War and Peace in the Upper South*. Baton Rouge: Louisiana State Univ. Press, 1988.

Babcock, Stephen, comp. *Babcock Genealogy*. New York: Eaton and Mains, 1903.

Bak, Richard. *A Distant Thunder: Michigan in the Civil War*. Ann Arbor, Mich.: Huron River, 2004.

Baker, Jean Harvey. *Mary Todd Lincoln: A Biography*. New York: W. W. Norton, 1989.

Ballard, Michael B. *Vicksburg: The Campaign That Opened the Mississippi*. Chapel Hill: Univ. of North Carolina Press, 2010.

Barnett, Le Roy, and Roger Rosentreter. *Michigan's Early Military Forces: A Roster and History of Troops Activated Prior to the American Civil War*. Detroit: Wayne State Univ. Press, 2003.

Belknap, Charles E. *History of the Michigan Organizations at Chickamauga, Chattanooga, and Missionary Ridge*. 2nd ed. Lansing, Mich.: Robert Smith, 1899.

Beth, Loren P. *John Marshall Harlan: The Last Whig Justice*. Lexington: Univ. Press of Kentucky, 1992.

Beyer, Walter H., and Oscar F. Keydel, eds. *Deeds of Valor*. . . . Vol. 1. Detroit: Perrien-Keydel, 1907.

Bickham, W. D. *Rosecrans' Campaign with the Fourteenth Army Corps, or the Army of the Cumberland: A Narrative of Personal Observations, with an Appendix, Consisting of Official Reports of the Battle of Stone River*. Cincinnati: Moore, Wilstach, Keys, 1863.

Bingham, S. D. *Early History of Michigan with Biographies of State Officers, Members of Congress, Judges and Legislators*. Lansing, Mich.: Thorp and Godfrey, 1888.

Boatner, Mark Mayo. *The Civil War Dictionary*. Rev. ed. New York: Vintage, 1991.

Bowling, W. K. *Cholera As It Appeared in Nashville in 1849, 1850, 1854 and 1866*. Nashville: Univ. Book and Job Office, Medical College, 1866.

Bowman, Virginia McDaniel. *Historic Williamson County: Old Homes and Sites*. Franklin, Tenn.: Sovran Bank, 1989.

Bradley, Michael R. *Tullahoma: The 1863 Campaign for the Control of Middle Tennessee*. Shippensburg, Pa.: Burd Street Press, 2000.

Brandt, Robert S. *Touring the Middle Tennessee Backroads*. Winston-Salem, NC: John F. Blair, 1995.

Broadwater, Robert P. *Civil War Medal of Honor Recipients: A Complete Illustrated Record*. Jefferson, N.C.: McFarland, 2007.

Busey, John W. *These Honored Dead: The Union Casualties at Gettysburg*. 2nd ed. Hightstown, N.J.: Longstreet House, 1996.

Castel, Albert. *Decision in the West: The Atlanta Campaign of 1864*. Lawrence: Univ. Press of Kansas, 1992.

Catton, Bruce. *This Hallowed Ground*. Ware, Hertfordshire: Wordsworth Editions, 1998.

Cist, Henry M. *The Army of the Cumberland*. New York: Charles Scribner's Sons, 1882.

Clearwater, Alphonso T., ed. *The History of Ulster County, New York*. Kingston, N.Y.: W. J. Van Deusen, 1907.

Coffin, Charles Carleton. *The Boys of '76: A History of the Battles of the Revolution*. New York: Harper and Bros., 1876.

Connelly, Thomas Lawrence. *Autumn of Glory: The Army of Tennessee, 1862–1865*. Baton Rouge: Louisiana State Univ. Press, 2001.

The Court Journal: From January to December 1835. London: W. Thomas, 1835.

Cowles, Albert E. *Past and Present of the City of Lansing and Ingham County, Michigan*. Lansing, Mich.: Michigan Historical, [1905?].

Cozzens, Peter. *No Better Place to Die: The Battle of Stones River*. Urbana: Univ. of Illinois Press, 1991.

———. *The Shipwreck of Their Hopes: The Battles for Chattanooga*. Urbana: Univ. of Illinois Press, 1996.

———. *This Terrible Sound: The Battle of Chickamauga*. Urbana: Univ. of Illinois Press, 1992.

Daniel, Larry J. *Battle of Stones River: The Forgotten Conflict between the Confederate Army of Tennessee and the Union Army of the Cumberland*. Baton Rouge: Louisiana State Univ. Press, 2012.

Daniel, Larry J., and Lynn N. Bock. *Island No. 10: Struggle for the Mississippi Valley.* Tuscaloosa: Univ. of Alabama Press, 1996.

Dempsey, Jack. *Michigan and the Civil War: A Great and Bloody Sacrifice.* Charleston, S.C.: History Press, 2011.

Devine, Shauna. *Learning from the Wounded: The Civil War and the Rise of American Medical Science.* Chapel Hill: Univ. of North Carolina Press, 2014.

Dickason, David Howard. *The Daring Young Men: The Story of the American Pre-Raphaelites.* New York: Benjamin Blom, 1970.

Donald, David Herbert. *Lincoln.* New York: Simon and Schuster, 1996.

Donald, David Herbert, Jean Harvey Baker, and Michael F. Holt. *The Civil War and Reconstruction.* New York: W. W. Norton, 2001.

DuBose, Joel C., ed. *Notable Men of Alabama: Personal and Genealogical, with Portraits.* Vol. 2. Atlanta: Southern Historical Association, 1904.

Duke, Basil W. *History of Morgan's Cavalry.* Cincinnati: Miami Printing, 1867.

Durant, Samuel W. *History of Ingham and Eaton Counties, Michigan, with Illustrations and Biographical Sketches of Their Prominent Men and Pioneers.* Philadelphia: D. W. Ensign, 1880.

Durham, Walter T. *Nashville: The Occupied City.* Knoxville: Univ. of Tennessee Press, 2008.

Engle, Stephen Douglas. *Don Carlos Buell: Most Promising of All.* Chapel Hill: Univ. of North Carolina Press, 1999.

Evans, E. Raymond. *The Battle of Davis' Crossroads: September 10–11, 1863.* Signal Mountain, Tenn.: CASI, 2008.

Fanebust, Wayne. *Major General Alexander M. McCook, USA: A Civil War Biography.* Jefferson, N.C.: McFarland, 2013.

Faust, Drew Gilpin. *This Republic of Suffering: Death and the American Civil War.* New York: Alfred A. Knopf, 2008.

Faust, Patricia L., ed. *Historical Times Illustrated Encyclopedia of the Civil War.* New York: HarperPerennial, 1991.

Fitch, John. *Annals of the Army of the Cumberland: Comprising Biographies, Descriptions of Departments, Accounts of Expeditions, Skirmishes, and Battles. . . .* 5th ed. Philadelphia: J. B. Lippincott, 1864.

Foner, Eric. *Reconstruction: America's Unfinished Revolution, 1863–1877.* New York: Perennial Classics, 2002.

Foote, Lorien. *The Gentlemen and the Roughs: Manhood, Honor, and Violence in the Union Army.* New York: New York Univ. Press, 2010.

Gallagher, Gary W. *The Union War.* Cambridge, Mass.: Harvard Univ. Press, 2011.

———, ed. *The Richmond Campaign of 1862: The Peninsula and the Seven Days.* Chapel Hill: Univ. of North Carolina Press, 2000.

Glatthaar, Joseph T. *Forged in Battle: The Civil War Alliance of Black Soldiers and White Officers.* New York: Free Press, 1990.

Gott, Kendall D. *Where the South Lost the War: An Analysis of the Fort Henry–Fort Donelson Campaign, February 1862.* Mechanicsburg, Pa.: Stackpole Books, 2003.

Gracie, Archibald. *The Truth about Chickamauga.* Boston: Houghton Mifflin, 1911.

Groom, Winston. *Shiloh, 1862.* Washington, D.C.: National Geographic Society, 2012.

Guelzo, Allen C. *Gettysburg: The Last Invasion*. New York: Alfred A. Knopf, 2013.

Guernsey, Alfred H., and Henry Mills Alden. *Harper's Pictorial History of the Great Rebellion*. 2 vols. New York: Harper, 1866–1868.

Hanna, Doreen Potter. *Kibbe Genealogical Notes: On Some Descendants of Edward Kibbe and His Wife Mary (Partridge) Kibbe*. 1972. Reprint, n.p.: iRoots.Net, 2000.

Harrell, Carolyn L. *When the Bells Tolled for Lincoln: Southern Reaction to the Assassination*. Macon, Ga.: Mercer Univ. Press, 1997.

Harrison, Bruce H. *The Family Forest Descendants of Edward III, King of England*. Kamuela, Hawaii: Millisecond, 2009. PDF e-book.

———. *The Family Forest Descendants of John Alden, Mayflower Pilgrim*. Kamuela, Hawaii: Millisecond, 2009. PDF e-book.

———. *The Family Forest Descendants of Thomas Blodgett*. Kamuela, Hawaii: Millisecond, 2009. PDF e-book.

Haynie, J. Henry, ed. *The Nineteenth Illinois: A Memoir of a Regiment of Volunteer Infantry Famous in the Civil War of Fifty Years Ago for Its Drill, Bravery, and Distinguished Services*. Chicago: M. A. Donahue, 1912.

Hebert, Walter H. *Fighting Joe Hooker*. Lincoln, Univ. of Nebraska Press, 1999.

Heiskell, S. G. *Andrew Jackson and Early Tennessee History Illustrated*. Vol. 3. Nashville: Ambrose, 1921.

Hess, Earl J. *Kennesaw Mountain: Sherman, Johnston, and the Atlanta Campaign*. Chapel Hill: Univ. of North Carolina Press, 2013.

———. *The Union Soldier in Battle: Enduring the Ordeal of Combat*. Lawrence: Univ. Press of Kansas, 1997.

Historical Catalogue of the Students of Kalamazoo College and of Kalamazoo Theological Seminary, 1851–1902. Kalamazoo, Mich.: Ihling Bros. and Everard, 1903.

History of Branch County, Michigan. Coldwater, Mich.: Branch County Historical Society, 1979.

History of Hillsdale County, Michigan, with Illustrations and Biographical Sketches of Some of Its Prominent Men and Pioneers. Philadelphia: Everts and Abbott, 1879.

The History of Jo Daviess County, Illinois, Containing a History of the County—Its Cities, Towns, Etc. . . . Chicago: H. F. Kett, 1878.

History of St. Joseph County, Michigan, with Illustrations Descriptive of Its Scenery, Palatial Residences, Public Buildings, Fine Blocks, and Important Manufactories, from Original Sketches by Artists of the Highest Ability. Philadelphia: L. H. Everts, 1877.

Hoobler, James A. *A Guide to Historic Nashville, Tennessee*. Charleston, S.C.: History Press, 2008.

———. *Nashville: From the Collection of Carl and Otto Giers*. Vol. 2. Charleston, S.C.: Arcadia, 2000.

Howard, Jerome Bird, ed. *Phonographic Magazine and National Shorthand Reporter*. 32 vols. Cincinnati: Phonographic Institute, 1887–1918.

Jenkins, William H., and John Knox. *The Story of Decatur, Alabama*. Decatur, Ala.: Decatur Printing, 1970.

Johnson, Mark W. *That Body of Brave Men: The U.S. Regular Infantry and the Civil War in the West*. Cambridge, Mass.: Da Capo, 2003.

Johnston, Henry P., ed. *The Record of Connecticut Men in the Military and Naval Service during the War of the Revolution, 1775–1783.* Hartford, Conn.: n.p., 1889.

Kautz, August V. *The Company Clerk: Showing How and When to Make Out All the Returns, Reports, Rolls, and Other Papers. . . .* Philadelphia: J. B. Lippincott, 1865.

———. *Customs of Service for Non-Commissioned Officers and Soldiers As Derived from Law and Regulations and Practised in the Army of the United States. . . .* Philadelphia: J. B. Lippincott, 1864.

Keifer, Joseph Warren. *Slavery and Four Years of War: A Political History of Slavery in the United States, Together with a Narrative of the Campaigns and Battles of the Civil War in Which the Author Took Part: 1861–1865.* Vol. 1. New York: G. P. Putnam's Sons, 1900.

Knott, Stephen F. *Secret and Sanctioned: Covert Operations and the American Presidency.* New York: Oxford Univ. Press, 1996.

Knox, John. *A History of Morgan County, Alabama.* Decatur, Ala.: Morgan County Board of Revenue and Control, 1967.

Lamers, William Mathias. *The Edge of Glory: A Biography of General William S. Rosecrans, U.S.A.* Baton Rouge: Louisiana Univ. Press, 1999.

Lansing and Its Yesterdays: A Compilation of the Historical Material Published in the Seventy-Fifth Anniversary Edition of the Lansing State Journal, January 1, 1930. Lansing, Mich.: State Journal Company, [1930?].

Lee, Dan. *Kentuckian in Blue: A Biography of Major General Lovell Harrison Rousseau.* Jefferson, N.C.: McFarland, 2010.

Leeke, Jim, ed. *A Hundred Days to Richmond: Ohio's "Hundred Days" Men in the Civil War.* Bloomington: Indiana Univ. Press, 1999.

Leepson, Marc. *Desperate Engagement: How a Little-Known Civil War Battle Saved Washington, D.C., and Changed the Course of American History.* New York: Thomas Dunne, 2007.

Leslie, Frank. *Frank Leslie's Illustrated History of the Civil War: The Most Important Events of the Conflict between the States. . . .* Edited by Louis Shepheard Moat. New York: Mrs. Frank Leslie, 1895.

Lewis, William Terrell. *Genealogy of the Lewis Family in America, from the Middle of the Seventeenth Century down to the Present Time.* Louisville, Ky.: Courier-Journal, 1893.

Life of James W. Jackson: The Alexandria Hero, the Slayer of Ellsworth, the First Martyr in the Cause of Southern Independence. . . . Richmond, [Va.]: West and Johnson, 1862.

Linderman, Gerald F. *Embattled Courage: The Experience of Combat in the American Civil War.* New York: Free Press, 1987.

Logue, Larry M., and Michael Barton. *The Civil War Veteran: A Historical Reader.* New York: New York Univ. Press, 2007.

Lossing, Benson J. *The Pictorial Field-Book of the Revolution, or, Illustrations, by Pen and Pencil, of the History, Biography, Scenery, Relics, and Traditions of the War for Independence.* Vol. 1. New York: Harper and Bros., 1860.

Mann, Wayne C. "The Road to Murfreesboro: The Eleventh Michigan Volunteer Infantry from Organization through Its First Battle." Master's thesis. Kalamazoo: Western Michigan Univ., 1963.

Manning, Chandra. *What This Cruel War Was Over: Soldiers, Slavery, and the Civil War*. New York: Alfred A. Knopf, 2007.

Map of Williamson County, Tennessee, From New and Actual Surveys. Philadelphia: D. G. Beers, 1878.

Marszalek, John F. *Sherman: A Soldier's Passion for Order*. New York: Free Press, 1993.

Marten, James. *Sing Not War: The Lives of Union and Confederate Veterans in Gilded Age America*. Chapel Hill: Univ. of North Carolina Press, 2011.

McPherson, James M. *Battle Cry of Freedom: The Civil War Era*. New York: Oxford Univ. Press, 2003.

———. *For Cause and Comrades: Why Men Fought in the Civil War*. New York: Oxford Univ. Press, 1997.

Millard, Candice. *Destiny of the Republic: A Tale of Madness, Medicine and the Murder of a President*. New York: Doubleday, 2011.

Millbrook, Minnie Dubbs. *A Study in Valor: Michigan Medal of Honor Winners in the Civil War*. [Lansing, Mich.]: Michigan Civil War Centennial Observance Commission, [1966].

Mitchell, Reid. *Civil War Soldiers*. New York: Viking, 1988.

Moore, Charles. *History of Michigan*. Vol. 3. Chicago: Lewis, 1915.

Morrison, George Austin. *King Genealogy: Clement King of Marshfield, Mass., 1668, and His Descendants*. Ltd. ed. Albany, N.Y.: Joel Munsell's Sons, 1898.

Nelson, Megan Kate. *Ruin Nation: Destruction and the American Civil War*. Athens: Univ. of Georgia Press, 2012.

Northern Alabama: Historical and Biographical Illustrated. Birmingham, Ala.: Smith and De Land, 1888.

Oden, Sue Barton. *Hold Us Not Boastful: A History of Thompson's Station, Tennessee and Its People*. Thompson's Station, Tenn.: Creative Designs, 1996.

Orton, Richard H., comp. *Records of California Men in the War of the Rebellion, 1861 to 1867*. Sacramento: J. D. Young, 1890.

Parton, James. *Life of Andrew Jackson*. Vol. 3. New York: Mason Bros., 1861.

Partridge, Charles A., ed. *History of the Ninety-Sixth Regiment Illinois Volunteer Infantry*. Chicago: [Brown, Pettibone], 1887.

Pearce, George F. *Pensacola during the Civil War: A Thorn in the Side of the Confederacy*. Gainesville: Univ. Press of Florida, 2008.

Pennypacker, Morton. *General Washington's Spies on Long Island and New York*. Brooklyn, N.Y.: Long Island Historical Society, 1939.

Pfanz, Donald. *Richard S. Ewell: A Soldier's Life*. Chapel Hill: Univ. of North Carolina Press, 1998.

Phillips, Margaret I. *The Governors of Tennessee*. 2nd ed. Gretna, La.: Pelican, 2001.

Portrait and Biographical Album of St. Joseph County, Michigan, Containing Full Page Portraits and Biographical Sketches of Prominent and Representative Citizens of the County. . . . Chicago: Chapman Bros., 1889.

Portrait and Biographical Record of Kalamazoo, Allegan and Van Buren Counties, Michigan, Containing Biographical Sketches of Prominent and Representative Citizens. . . . Chicago: Chapman Bros., 1892.

Powell, David A. *The Maps of Chickamauga: An Atlas of the Chickamauga Campaign, Including the Tullahoma Operations, June 22–September 23, 1863*. Cartography by David A. Friedrichs. New York: Savas Beatie, 2009.

Powell, Lawrence N. *New Masters: Northern Planters during the Civil War and Reconstruction*. 1980. Revised with new preface. New York: Fordham Univ. Press, 1998.

Proceedings of the Michigan Press Association at the Fifteenth Annual Meeting. . . . Greenville, Mich.: Independent Steam, 1882.

Quinlan, James Eldridge. *History of Sullivan County. . . .* Liberty, N.Y.: G. M. Beebe and W. T. Morgans, 1873.

Rable, George C. *Fredericksburg! Fredericksburg!* Chapel Hill: Univ. of North Carolina Press, 2002.

Ramage, James A. *Rebel Raider: The Life of General John Hunt Morgan*. Lexington: Univ. Press of Kentucky, 1995.

Ramold, Steven J. *Baring the Iron Hand: Discipline in the Union Army*. DeKalb: Northern Illinois Univ. Press, 2010.

Record of Service of Michigan Volunteers in the Civil War, 1861–1865. 46 vols. Kalamazoo, Mich.: Ihling Bros. and Everard, [1905?].

Rexford, John Dewitt, comp. *Genealogical History Showing the Paternal Line of Descent from Arthur Rexford, a Native of England, Who Married Elizabeth Stevens, of New Haven, Conn., in 1702*. Janesville, Wis.: Gazette, 1891.

Riker, James, and Henry Pennington Toler. *Revised History of Harlem (City of New York): Its Origin and Early Annals. . . .* Edited by Sterling Potter. New York: New Harlem, 1904.

Roberts, Gary Boyd. *The Royal Descents of 600 Immigrants to the American Colonies or the United States Who Were Themselves Notable or Left Descendants Notable in American History*. Vol. 1. Baltimore: Genealogical, 2008.

Roberts, James A. *New York in the Revolution As Colony and State*. Albany, N.Y.: Weed-Parsons, 1897.

Robertson, John, comp. *Michigan in the War*. Rev. ed. Lansing, Mich.: W. S. George, 1882.

Roll of State Officers and Members of General Assembly of Connecticut from 1776 to 1781. . . . Hartford, Conn.: Case, Lockwood and Brainard, 1881.

Rothman, Ellen K. *Hands and Hearts: A History of Courtship in America*. New York: Basic Books, 1984.

Salling, Stuart. *Louisianians in the Western Confederacy: The Adams-Gibson Brigade in the Civil War*. Jefferson, N.C.: McFarland, 2010.

Savage, John. *The Life and Public Services of Andrew Johnson, Seventeenth President of the United States, Including His State Papers, Speeches and Addresses*. New York: Derby and Miller, 1866.

Sears, Stephen W. *Chancellorsville*. New York: Houghton Mifflin, 1996.

———. *George B. McClellan: The Young Napoleon*. New York: Da Capo, 1999.

———. *Landscape Turned Red: The Battle of Antietam*. New York: Houghton Mifflin, 2003.

———. *To the Gates of Richmond: The Peninsula Campaign*. New York: Ticknor and Fields, 1992.

Shenstone, Susan Burgess. *So Obstinately Loyal: James Moody, 1744–1809.* Montreal: McGill-Queen's Univ. Press, 2002.

Silliman, Sue I. *St. Joseph in Homespun: A Centennial Souvenir.* Three Rivers, Mich.: Three Rivers Pub., 1931.

Smith, Jean Edward. *Grant.* New York: Simon and Schuster, 2001.

Smith, Timothy B. *A Chickamauga Memorial: The Establishment of America's First Civil War National Military Park.* Knoxville: Univ. of Tennessee Press, 2009.

Speer, William S. *Sketches of Prominent Tennesseans, Containing Biographies and Records of Many of the Families Who Have Attained Prominence in Tennessee.* Nashville: Albert B. Tavel, 1888.

Stahura, Barbara. *Sons of Union Veterans of the Civil War.* Compiled and edited by Gary L. Gibson. Paducah, Ky.: Turner, 1996.

Stevens, Charles A. *Berdan's United States Sharpshooters in the Army of the Potomac.* St. Paul, Minn.: Price-McGill, 1892.

Stevenson, Alexander F. *The Battle of Stone's River Near Murfreesboro', Tenn., December 30, 1862, to January 3, 1863.* Boston: James R. Osgood, 1884.

Stilwell, E. J. *History of Company K, 33d Michigan Volunteer Infantry, U.S.A.* Constantine, Mich.: Clemens Bros., 1899.

Storey, Margaret M. *Loyalty and Loss: Alabama's Unionists in the Civil War and Reconstruction.* Baton Rouge: Louisiana State Univ. Press, 2004.

Thornton, Leland W. *When Gallantry Was Commonplace: The History of the Michigan Eleventh Volunteer Infantry, 1861–1864.* New York: Peter Lang, 1991.

Three Rivers: The Early Years. Three Rivers, Mich.: Three Rivers Sesquicentennial Committee, 1986.

Tomlinson, Everett T. *Stories of the American Revolution.* Part 1. Boston: Thomas R. Sherwell, 1898.

Trelease, Allen W. *White Terror: The Ku Klux Klan Conspiracy and Southern Reconstruction.* Baton Rouge: Louisiana State Univ. Press, 1971.

Trudeau, Noah Andre. *The Last Citadel: Petersburg, Virginia, June 1864–April 1865.* Baton Rouge: Louisiana State Univ. Press, 1993.

Weber, Jennifer L. *Copperheads: The Rise and Fall of Lincoln's Opponents in the North.* New York: Oxford Univ. Press, 2008.

Wells, Clayton B. *My Father Benjamin F. Wells.* Ann Arbor, Mich.: Alumni Press, 1929.

Wheelan, Joseph. *Terrible Swift Sword: The Life of General Philip H. Sheridan.* New York: Da Capo, 2012.

Wiley, Bell Irvin. *The Life of Billy Yank: The Common Soldier of the Union.* Baton Rouge: Louisiana State Univ. Press, 1971.

Wills, Brian Steel. *George Henry Thomas: As True as Steel.* Lawrence: Univ. Press of Kansas, 2012.

Woodworth, Steven E. *While God Is Marching On: The Religious World of Civil War Soldiers.* Lawrence: Univ. Press of Kansas, 2001.

Articles, Papers, and Online Sources

Casey, William J. "Remarks of William J. Casey, Director of Central Intelligence, Before the Sons of the American Revolution." Paper presented in Palm Beach, Fla., February 21, 1983.

Coleman, Ellen Snell. "The Harding House." *Rutherford County Historical Society*, no. 23 (Summer 1984): 1–13.

Hershock, Martin J. "Copperheads and Radicals: Michigan Partisan Politics during the Civil War Era." *Michigan Historical Review*, Spring 1992, 29–69.

Hill, Kendrick C. "National Stenographer's Association." *Stenographer* 13, no. 4 (April 1898): 77–79.

"James W. King." *Student's Journal* 18, no. 7 (July 1889): 1–3.

Lewis, Jim. "The Battle of Stones River." *Blue & Gray* 28, no. 6 (2012): 6–50.

"The Mail from New-York." *London Political Magazine and Parliamentary, Naval, Military, and Literary Journal*, June 1781, 343–44.

Robertson, William Glenn. "The Chickamauga Campaign, McLemore's Cove: Rosecrans' Gamble, Bragg's Lost Opportunity." *Blue & Gray*, Spring 2007, 6–50.

———. "The Chickamauga Campaign, The Battle of Chickamauga: Day 2, September 20, 1863." *Blue & Gray*, Summer 2008, 6–50.

Smith, Elizabeth Oakes. "The Intercepted Messenger of Ramapo Pass." *United States Magazine*, no. 3 (July–December 1856): 115–25, 225–35.

United States Census. "U.S. Federal Census Collection." Last accessed February 20, 2015. http://search.ancestry.com/search/group/usfedcen.

Index